CHRIS BOARDMAN

THE BIOGRAPHY OF THE
BIKE

First published in 2015 by Voyageur Press, an imprint of Quarto Publishing Group
USA Inc., 400 First Avenue North, Suite 400, Minneapolis, MN 55401 USA

Published by arrangement with Octopus Publishing Group Ltd, Carmelite House,
50 Victoria Embankment, London EC4Y 0DZ

Voyageur Press titles are also available at discounts in bulk quantity for industrial or
sales-promotional use. For details write to Special Sales Manager at Quarto Publishing
Group USA Inc., 400 First Avenue North, Suite 400, Minneapolis, MN 55401 USA.

To find out more about our books, visit us online at www.voyageurpress.com.

ISBN: 978-0-7603-4989-2

Printed and bound in China

Publisher Trevor Davies
Senior Editor Leanne Bryan
Copy-editor Joanna Chisholm
Proofreader Miranda Harrison
Indexer Helen Snaith
Deputy Art Director Yasia Williams-Leedham
Designer Grade Design www.gradedesign.com
Special Photography Tif Hunter
Photographic Assistant Jack Fillery
Picture Research Manager Giulia Hetherington
Picture Researcher Nick Wheldon
Senior Production Manager Peter Hunt

CHRIS BOARDMAN

THE BIOGRAPHY OF THE
BIKE

With Chris Sidwells

Voyageur
Press

Contents >

Introduction >

Bikes were part of my childhood, like they are for most kids, but my mother and father were keen cyclists and my father raced successfully and still rides. Neither pushed me into racing, but bikes were inevitably a bigger part of my early life than they were for most of my friends. Except one, Scott O'Brien. He had cycling parents, too, and he started racing. So although I enjoyed other sports I found myself cycling more, then I raced, and something bit.

My first race was a 10–mile time trial. At the end of a time trial you get a number; your time. The next week on the same course I went faster and lowered my number, and I liked it. I liked that there was something quantifiable to work with – and lowering the number, getting faster, spurred me on.

My racing career was an experiment to see how fast I could go, and in carrying out that experiment I became fascinated with bikes and how they worked. I studied bikes, tried to understand the interaction of the human body with the bike and the biomechanics of the whole package. With what I learned – and with highly skilled people helping me – I began to see the bike and the improvements that could be made to it as tools I could use to help me ride faster.

I kept that approach throughout my racing career, being open to innovation no matter how against the grain of tradition it might have appeared. I tried new ideas, keeping those that worked and discarding what didn't. I use a similar approach – although with less trial and error due to modern computer modelling – to produce and develop my own bike brand. I also brought this approach to my work with bikes and materials as part of British Cycling's performance effort.

Now, though, I see bikes as more than just a tool, more than something to be refined and developed in search of speed. I see bikes as having a part to play in wider society. Bikes are one of the answers to congestion in our towns and cities. Cycling can help people get fitter and healthier and help them live longer and fuller lives. Bikes can help reduce pollution all over the planet. They also have a big role to play in developing countries.

So that's the story of my life and my relationship to bikes so far. In this book I've tried to tell the story of the bike, how it has developed over time. I've also tried to show how bikes have made a difference to the world we live in. This is the biography of the bike; its birth, its development, its present and its future.

RIGHT The author in his role as Technical Advisor to the Great Britain Cycling Team at the Manchester Velodrome in March 2009.

ABOVE The author on his way to winning the first British cycling gold medal for 72 years in the Barcelona 1992 Olympic Games. The event was the 4000m individual pursuit.

LEFT A wind tunnel. A huge amount of research has been done using these tubular passages to improve and refine the aerodynamics of cycling, in terms of both bike construction and rider position.

Chapter 1
The First Bikes >

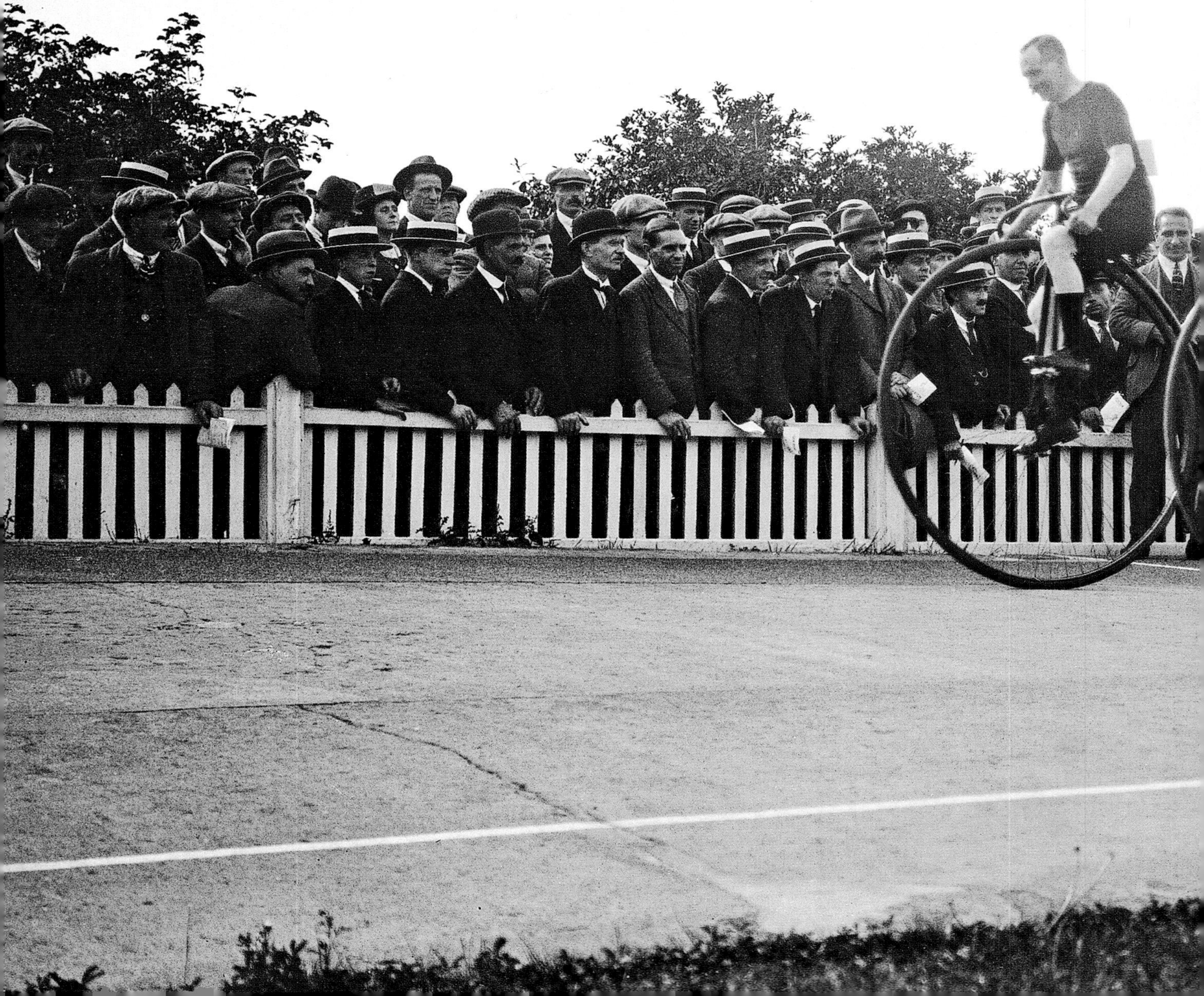

Contestants in the 1919 Challenge
High Ordinary race organized by the
National Cycling Union at Herne Hill
in South London.

Inventing Cycling >

Pedalling is the thing that defines cycling, so a bicycle is only a bicycle when it is being pedalled along the road or trail. Human-powered land vehicles, some with two wheels, others with more, existed for a long time – the earliest recorded being built by Giovanni Fontana in 1418. However, these machines only became bicycles in the mid-19th century, when pedals were attached to them.

Pedalling in a form that we would recognize today was invented in 1864 by a Paris blacksmith named Pierre Michaux, who attached pedals and cranks to the front wheel of a two-wheeled machine called a draisine, which was named after its inventor Baron Drais von Sauerbronn.

In 1817 von Sauerbronn had taken a design that emerged in the second half of the 18th century and improved it in several ways. Most importantly, he had added a dependable easy way to steer the front wheel. Riders sat astride a draisine on a very basic saddle, held its rudimentary handlebars and scooted themselves along the ground with their feet. As it was a running motion, it was called a Laufmaschine (running machine) in Germany.

Von Sauerbronn took out a patent in 1818 in the UK, where a coach maker called Denis Johnson acted for him. British users rechristened the baron's invention the hobby horse then the dandy horse, because draisines became very fashionable with Regency gentlemen. However, draisine riding also became a nuisance. Because these riders preferred smooth pavements to rough roads, they tended to travel on pavements and on park footpaths, where inevitable collisions with pedestrians resulted in such practices being outlawed. The dandy horse went out of fashion nearly as quickly as it came in.

Improving the draisine >

Meanwhile, other inventors looked at the design of the dandy horse to see if it could be improved. In the 1890s, the family of Kirkpatrick Macmillan, a Scottish blacksmith, claimed that he had invented a treadle-driven bicycle back in 1839 and that it broke foot contact with the floor, but there is insufficient evidence to back up this claim. Michaux's invention of the 1860s is much better documented.

A customer brought Michaux a French version of the draisine and asked to have some footrests placed where the front-wheel hub was attached to the frame, so that he could coast downhill without his feet touching the floor and travel faster. Michaux took this idea and developed it. Why not, as with the powered grindstone in his workshop, drive the front wheel directly with your legs and feet?

Michaux's thought was the creative spark that led directly to the invention of the modern bike.

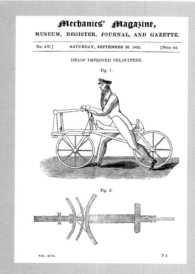

LEFT A later draisine (or hobby horse as they were called in the UK) with metal as well as wooden parts. Early models were all wood. This one has a padded leather chest rest that the rider could lean against to gain some extra push.

RIGHT An illustration of an early wooden draisine, showing how the rider propels it along with his feet.

As the 1860s drew to a close, the Oliviers were the world's biggest bicycle manufacturers but, in 1870, the Franco-Prussian War drastically slowed their progress, handing the lead in bicycle development to the British.

The velocipede >

Michaux adapted his customer's bike and, with the idea proving to work, he adapted others. Emboldened, Michaux and his partner, Pierre Lallement, began manufacturing their own version of the draisine, with pedals, and called it a velocipede. This name had originally been coined by Nicéphore Niépce for his earlier improvement of a draisine, but his design did not have pedals. The shortened form, velo, is the French for bike.

Michaux and Lallement took out a patent in 1866, and by 1869 there were 60 velocipide manufacturers in Paris and 15 more around France. Michaux and Lallement also took out a patent for the velocipede in America.

Big business >

Once bicycle manufacturing was established, the biggest company in the early days was the Compagnie Parisienne des Vélocipèdes, which was owned and run by two brothers, René and Aimé Olivier. They claimed to deliver 200 velocipides a day from their factory in the rue Bugeaud. As well as being good businessmen, the Oliviers were cycling enthusiasts and as such they played a big part in the next stage of the bike's story – the development of cycle racing.

The Oliviers were one of the first manufacturers to produce bikes that were made from metal. They also backed one of the first ever cycle shows, held at Le Pré Catelan in Paris, 1–5 November 1869.

As the 1860s drew to a close, the Oliviers were the world's biggest bicycle manufacturers but, in 1870, the Franco-Prussian War drastically slowed their progress. It was at this point that the British took the lead in bicycle development. This did not affect the Oliviers' wealth too much. They came from a rich family who owned a chemical plant in Lyon and, although they were passionate about cycling itself, their bike company was more a reflection of their interest than a commercial necessity.

RIGHT An early cycle race, in the Paris suburb of Levallois, promoted by the Compagnie Parisienne. The Vélo Club Levallois was formed in 1891 and is one of the oldest cycling clubs in France.

LEVALLOIS, près PARIS. — Course de Vélocipèdes avec obstacles. Nouveau système de vélocipèdes fabriqués par la Compagnie parisienne.

Boneshaker, the First Mass-Produced Bike >

Pierre Michaux and Pierre Lallement's first bike was made entirely of wood. Later, they added metal tyres to stop the wheels wearing out but that did not do anything to cut the harshness of riding a velocipede on cobbled roads. Because of the harsh ride, by the time they went into mass production, the velocipede was also known as the boneshaker.

Production advanced. Wrought-iron frames were made, still with wooden wheels paired to metal tyres. To make cycling a little safer, Michaux added a spoon brake to his first mass-produced bikes (see page 120). Early versions of these brakes acted on the rear wheel, by pressing a wooden pad directly on to the metal tyre. The pad was applied by pulling a chord that ran between the brake and the handlebars.

The saddles of early metal frames were mounted on a long flat spring, which ran from the frame's main structural element to the rear wheel to give them an element of suspension. By 1870, technical advances in metallurgy and other materials resulted in the manufacture of the first all-metal bikes. These also had solid rubber tyres.

The rear hubs of these early bikes ran with metal on metal, with little lubrication, while the front hubs, which had to tolerate torque forces from pedalling, had lubricated bronze bearings. Some front hubs even had lubrication tanks attached to them, from which oil seeped down into the hub from a reservoir of lamb's wool that was soaked in oil.

RIGHT/OPPOSITE A wrought-iron boneshaker. Note the rudimentary metal leaf spring on which the metal saddle is attached. The spoon brake acts on the rear wheel and is operated by pulling a control wire attached to the handlebars [1]. Detail of the thick round wooden spokes, the rim and the metal tyre [2]. The flat platform brass pedals have an oil port, which could be adjusted within the metal cranks to suit the rider's inside leg measurement [3].

The birth of cycling media >

Not long after cycling was invented, cycling magazines were introduced to talk about it. *Le Vélocipède Illustré* was one of the first. Published in Paris, issue number one, dated 1 April 1869, showed an engraving of a woman riding a Michaux velocipede whilst carrying a flag bearing the word 'Progress'.

The last line of the first editorial, written by Richard Lesclide under his pen name Le Grand Jacques, read: 'The bicycle has now won complete acceptance in France, and we are founding a magazine under its patronage in order to bring together, in the same fellowship, its adherents and believers.'

Much like most cycling magazine editors around the world today, Lesclide was a cyclist and a believer in cycling. But just like all non-state-supported media, *Le Vélocipède Illustré* had to pay its way through circulation and advertising. Its publication saw the beginning of cycling depicted in art under the guise of advertisements.

Vélocipède Michaux 1869
Don de MM^{rs} Olivier de Sanderval et Frédéric Dumesnil

COMPAGNIE PARISIENNE

ABOVE An elegant Michaux velocipede showing the pulley refinements to the spoon brake that would have increased the force with which the brake could be applied.

RIGHT A Player's cigarette card, celebrating the Michaux velocipede.

PLAYER'S CIGARETTES

MICHAUX VELOCIPEDE

2

Ready, Steady, Go! >

Cycling was born in France, so it is natural that cycle racing was born there too. Cycling must have appealed to something in the French psyche, because the delight of moving great distances or at speed under one's own power on this new machine spread quicker in France than it did elsewhere in Europe. Within just a few years of Pierre Michaux's invention, there were indoor cycling academies all over Paris. French parks were full of young men and women riding together. More adventurous male soloists rode fast and did two-wheeled tricks. It was not long before somebody said: 'Let's have a race.'

No doubt there were races before and there were even others on the same day, but one race is so well documented that it is accepted as being the first ever. Organized by the Olivier brothers – the Compagnie Parisienne des Vélocipèdes manufacturers – the race was held in the Parc St Cloud, Paris, on 31 May 1868 and was won by a young Englishman named James Moore.

Moore was born in Suffolk and moved to Paris with his parents when he was young – living, in fact, on the same street where Michaux had his workshop. Moore was fascinated by Michaux's invention and had a velocipede from the age of 16. He even learned to do tricks on it, having been taught these by acrobats from the nearby Cirque d'Été.

The race that Moore won was 1km (0.6 miles) long and his time for the distance was 2 minutes 35 seconds, giving him an average speed of 23.2km/h (14.4mph). There were four other contestants and Moore's race was only one of several that took place in the Parc St Cloud that day. However, some of them were novelty races – such as the slow bicycle race – and others were for smaller-wheeled bikes, so they would not have been as fast as Moore's event.

LEFT The Compagnie Parisienne, owned by brothers Aimé and René Olivier, did a great deal to promote the first cycle races. Avid touring cyclists, Aimé and René had many adventures on two wheels. In August 1865 the brothers rode from Paris to Avignon with a journalist friend, Georges de la Bouglise. It took them eight days to cover 794km (493 miles) on Michaux Vélocipèdes. The ride is considered to be the first ever touring expedition on bicycles.

Paris to Rouen >

The desire to race bikes spread through Europe after the Parc St Cloud event. The first race held in Britain was the very next day – 1 June 1868, followed on 18 July by an event in Ghent, Belgium, and another in September in Brno in Morovia (the current Czech Republic). But it was in the following year (1869) that cycle racing really took root and people began wondering what could be achieved on this new invention.

John Mayall junior set things rolling that year when he covered the 83km (51 miles) from London to Brighton in around 12 hours. Yet Mayall's was a personal challenge; for the first official place-to-place race we have to return to France.

It happened on 7 November 1869, and the route was 123km (76 miles) from Paris to Rouen. Some 120 contestants started, including two women – one of whom gave her name as Miss America, presumably because she wanted to keep her participation a secret. Thirty-four riders made it to Rouen within the cut-off time of 24 hours, including Miss America in 29th place. Her real identity has never been discovered.

James Moore won the Paris–Rouen event with a time of 10 hours 45 minutes, with an average speed of around 14km/h (9mph), in drizzly weather on cobblestones and dirt roads. His prize was 1,000 gold francs. It was the beginning of Moore's career as a professional cyclist. He went on to win many more races after Paris–Rouen and, in 1870, he set the first world hour record with a distance of 23.2km (14½ miles) on a cycle track in Wolverhampton.

ABOVE James Moore, on the right, with fellow competitor Jean-Eugène Castera before the start of the first Paris to Rouen road race in 1869.

Big Wheel Turning >

Racing has driven bike design more than any other factor and this was first evident in the front wheels of early bikes. By attaching pedals directly to a wheel, engineers created what is called a 1:1 drive, in which one revolution of the pedals equals one revolution of the wheel. The top speed was quickly limited by the pedal rate or cadence people could maintain. The way to go faster was to increase the diameter of the wheel being driven, so the bicycle covered more distance for each revolution.

When James Moore won his race in Parc St Cloud in 1868 the wheels on his bike were of roughly equal size. When he won the Paris–Rouen race in 1869 his bike's front wheel had a diameter of 48.25in and the rear had shrunk to 15.75in. With direct-drive it is only necessary to increase the size of the driven wheel to enhance speed; the other wheel has to follow whatever its size, so it may as well be small to save weight.

Moore's winning bike was also lighter than Michaux's first design in 1864 (see pages 10–11) – 25kg (55lb) compared to 30kg (66lb). It had solid rubber tyres that were attached to the wheel rims by a system patented by Clément Ader in 1868. Ader, who was an avid inventor, would later go on to be one of the pioneers of aviation.

The big wheel/little wheel look of Moore's machine was the blueprint for a type of bike that became known around the world as the Ordinary. In the UK it is called a penny farthing due to the disparity in wheel size, which reminded people of the different sizes of those two old coins. However, in its time the machine was referred to simply as a bicycle and, a little later, as a high-wheeler.

Not for road racing >

Although they were reasonably fast, high-wheeled bikes were not ideal for racing on roads and in its early days road racing was not much of a spectacle to watch. Racing around oval cinder tracks on high-wheelers, however, was. Moore took to track racing as well as he had taken to the road. In 1874, among many other races, he almost won the 0.6-km (1-mile) race in Wolverhampton, an event that was billed as the first ever world cycling championships.

RIGHT Known as the penny farthing in the UK, this ordinary illustrates how, in the pursuit of speed, the front wheel was made so much bigger than the back wheel. The bike also has wire-spoked metal wheels, solid rubber tyres and a leather saddle with springs between it and the bike's frame, to provide comfort.

That was the same year as the first Oxford versus Cambridge cycling match. The students taking part must have been very tall, because their bikes measured 1.5m (5ft) from ground to saddle. They beat the top French pro racer of the day, Camille Thulliet, whose bike was only 1.2m (4ft) high at a race meeting in Sheffield. The Hon. Ion Keith-Falconer of Cambridge raced on the highest bike, which had a 60-in diameter front wheel. However, he was still beaten at London's Lillie Bridge track in a 40-km (25-mile) race in 1876 by H P Whiting, who used a 54-in front wheel, in a time of 1 hour 41 minutes 16 seconds – although a 54-in wheel is hardly small.

Racing on high-wheeled bikes made for spectacular viewing and track meetings became ever more popular. (A small group of enthusiasts continue to hold track events for high-wheeled machines to this day.) High-wheelers, almost by definition, were inherently dangerous forms of transport. From a racing perspective, their speed potential, ostensibly dictated by the length of the rider, was also at its limit. Another way had to be found, meaning Whiting and Falconer's inseam advantage was nearing the end.

Quantifying ratios >

If a racing cyclist were to say: 'I used the 53 x 15 to attack at that moment of the race', the cyclist would be referring to a gear ratio of a 53-tooth chain-ring combined with a 15-tooth sprocket on the rear wheel. Due to the racing wheel size having been standardized for several decades, racing colleagues would understand what this short-hand description meant in terms of feel, effort and speed. But to truly know what gear a bicycle has, wheel diameter has to be factored in. The true measure of a gear is expressed as the distance a bike travels in one revolution of the pedals.

Europeans use a system called the gear ratio's development, which is derived using the formula: 'development = the drive wheel's circumference in metres multiplied by the number of teeth on the chain wheel divided by the number of teeth on the sprocket'. Development is how far the bike travels along the ground in one revolution of the pedals.

But American and UK bike experts prefer to stick with an older term: 'gear ratio = drive wheel diameter multiplied by the number of teeth on the chain-ring divided by the number of teeth on the sprocket'. The resulting gear number in inches is the distance the bike will travel in a single pedal revolution. (For example, a 26-in wheel being driven by a 42-tooth chain-ring and 18-tooth sprocket would give a ratio of 26 x 42/18 = 60.66, which is slightly bigger than the diameter of Falconer's 60" front wheel. The high-wheeler pedals were attached directly to the front wheel, so their equivalent of this number is simply the diameter of the front wheel.)

LEFT Ordinary racing on cinder tracks hit its heyday in the 1870s, but enthusiasts still race on these old bikes. This group in 1937 are training for the annual penny farthing race in London.

Danger Danger! >

Once the rider was mounted, a high-wheeler was not too difficult to ride on a smooth track and the large front wheel rolled well over small bumps. The problem was the rider's high centre of gravity and weight distribution, which was very far forwards compared to modern bicycles. If the front wheel hit a big bump in the road, the sudden loss of speed could pitch the rider over the handlebars to land head first. This kind of fall became known as 'taking a header'.

Riders used a bar welded to the bike's main frame to help them mount. With one foot up they would scoot themselves along while holding the handlebars until enough momentum was gathered. At this point they would step up and swing their 'scooting' leg over the saddle, before placing both feet on the pedals.

A rider also had to be nimble getting off. Skilful high-wheelers could balance when stationary, or almost stationary, by turning the front wheel and using forward and backward pressure on the pedals as well as their upper bodies as counterweights – like Fixie riders do at city traffic lights today. It was precarious, as any resulting fall was magnified by height.

Preventing headers >

Many things were tried to prevent headers. Catching knees on handlebars was a big cause of falls. To prevent this, Whatton handlebars were invented, which wrapped around the back of the rider's legs, and W-shaped handlebars, which gave rise to the term 'handlebar moustache', gave a rider's knees extra clearance. Towards the end of the high-wheeler's time, US designer G W Pressey even tried reversing the wheels to give more stability in the event of sudden slowing. Although this helped in the forward direction, the weight redistribution saw the rider topple over backwards when riding up steep hills.

Braking also had to be done very carefully. A simple spoon brake, operated from the handlebars by a rod and lever, acted on the front tyre, but applying it too hard could also cause a header.

To go downhill riders took their feet off the pedals and let the bike fly, but increased speed enhanced the danger. Many riders coasting downhill rested their legs on top of the handlebars, so if they were pitched off at least they landed feet first.

Skilled riders could prevent a header before the point of no return by riding for a short way with the back wheel off the ground. All in all, high-wheelers were not very practical and their reign was short. In fact, headers were so common that they may have been inadvertently responsible for the birth of the unicycle.

The invention of the unicycle cannot be attributed to any one person. There are pictures from the late 19th century showing cyclists riding with their rear wheels lifted and its thought that one of these riders simply cut off the back half of their bike.

High-wheelers gave cycling one thing that is still being used today – the wire-spoked wheel. Overall though, their biggest contribution to cycling was the need to find something better. It did not take long for inventors to come up with an improvement.

RIGHT Ordinary riders could take headers for all sorts of reasons, including applying the brake too hard or simply by going downhill. It was also possible that the rutted roads prevalent at the time caused tyres to roll off their rims. This advertisement from 1887 is for the 'Perfect Process', which claimed to be a better way of securing rubber tyres to bicycles and carriages.

FAR RIGHT The simplest way to mount an ordinary is to push off with one foot while placing the other on a pedal. When the front wheel rolls forwards, the crank and pedal attached directly to it lift the rider up into the saddle. Here, a more modern ordinary rider, George Rutland, demonstrates the process in 1964.

Wheels made using the tension of wire spokes are found on the majority of bikes today. Machines do the job now, but for a long time wire-spoked wheels were hand built, a skill that required training and practice.

The wheel-builder's art >

Eugene Meyer was the leading manufacturer of high-wheeled bikes in France and, in 1869, he patented a 'wire-spoked tension wheel'. The wheels became known as spider wheels in the UK, where James Starley of Coventry adapted them for his famous high-wheeler – the Ariel. The Ariel had a tangentially spoked front wheel, whereas Meyer's spokes were arranged in a radial pattern. Tangential spokes cross over each other, which increases the wheel's strength and helps absorb shock.

Wheels that are made using the tension of wire spokes can be found on the majority of bikes today. Machines do the job now, but for a long time wire-spoked wheels were hand built, a skill that required extensive training and practice. Specialist wheel builders raised their work to an art form when they made wheels for racing bikes.

To save weight and to improve aerodynamics (as the top of the wheel is travelling twice the forward speed of the bike), the number of spokes in a racing wheel was reduced, which meant that the quality of construction had to be perfect. The best wheel builders tensioned spokes as if they were tuning a musical instrument, by feeling and listening until the tension was pitch perfect.

Top wheel builders also played with tangential and radial spoke patterns, using tangential where strength was needed and radial spokes elsewhere, for lightness. Radial spokes are shorter and therefore lighter than the equivalent used in a tangential

spoked wheel. In later years, flat spokes, still essentially tensioned wire, were used to reduce air resistance even further.

Most wire-spoked wheels are built by inserting the spokes through holes in the flange of a hub. The spokes are then joined to the rim through threaded nipples. This part of the process is called lacing the wheel. Tension, and therefore the wheel's strength, is created by tightening the nipples, each a little in turn and continuing around the wheel until it runs strong and true.

ABOVE/LEFT Wheel building is a skill that requires training and lots of practice. Some wheels are still built by hand but most are made by machines. However, even the best made wheels can go out of line if a spoke breaks or they are knocked, so a good bike mechanic must still be able to adjust them to run true. Here, mechanics from the French wheel and bicycle equipment manufacturer Mavic demonstrate that skill.

An early cycle lamp advertisement, illustrating two Lucas battery-powered lights.

Borrowing from Industry >

Victorian times saw major leaps in engineering and technology. Most were connected with improving the productivity of industry, although there were many things that enhanced life in the home. A few of them were crucial in the story of the bike, facilitating a huge leap forward in its development.

Chain drive >

The roller chain had been around for a while. Engineer, inventor and industrialist Hans Renold is credited with inventing it in 1880, but there are sketches by Leonardo da Vinci as far back as the 16th century showing how this innovation would work.

The roller chain became increasingly used in Victorian machines, replacing belt drives, especially where space was constricted. Unlike with belts, there is no slippage with a chain drive, so where a lot of torque goes through the drive wheel, a chain transfers power more efficiently and effectively.

Although there have been different types of bicycle power transfer, such as belt drive, shaft drive and the Simpson lever chain, nothing has yet beaten the roller chain design for efficiency. Even today the chains on the most advanced bikes still follow the same principles as those that helped solve the shortcomings of the high-wheeled bike.

Two types of links make up a roller chain: inners and outers. Inner links are two flat plates of metal with a hole through each end, separated by a metal bush. Outer links are also flat plates of metal with holes at either end, but they fit either side of the inner links. A pin goes through the hole in the outer plates, through the inners and through each bush to hold them together. Early chains did not have bushes – they were added when Hans Renold invented the bush roller chain in 1880 to reduce friction and improve efficiency.

Ball bearings >

Ball bearings reduce rotational friction and support radial and axial loads. Before the introduction of ball bearings, a machine's rotating parts ran on direct contact bearings. Although lubrication and smooth metals, such as brass, reduced friction, direct bearings were not efficient and they wore down very quickly.

The ball bearing, or to be precise the radial ball bearing, is something cycling gave to industry, not the other way around. The first ever patent for a ball bearing was taken out by a British iron producer called Philip Vaughan. The first radial ball bearing – where balls are arranged around an axle that is fixed, and the hub revolves on the metal balls – was invented by Parisian bike mechanic, Jules Suriray. They were fitted to the Tribout bike ridden by James Moore when he won the Paris–Rouen race in 1869.

Despite that early success and their use on some later high-wheelers, ball bearings did not come into general use in the cycle industry until the next revolutionary bike was developed – the safety bicycle.

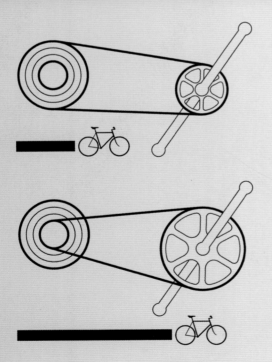

Gearing >

Simple gearing means rotating a larger toothed cog to turn a smaller toothed cog at a faster rate. The teeth of the cogs intermesh (or mesh, as engineers call it). One cog turning another is a basic machine and the process is called transmission.

Cogs do not have to mesh directly; they can be linked by a chain. On a bike, the larger cog, called a chain-ring, is rotated by pedalling. The chain-ring is linked to a smaller cog, called a sprocket, by a continuous roller chain.

Chain drive ended the development of high-wheeled bikes because there was no need for a bigger wheel on a chain-drive bike. A few were adapted with pedals attached to a chain-ring instead of directly to the front wheel. The rider's legs powered the chain-ring, which rotated a smaller sprocket attached to the larger wheel through a roller chain.

This arrangement meant that the front wheel could now travel faster than a rider's legs. But the higher speeds made high-wheelers even more dangerous and designers soon realized that chain-driven gears meant there was no need for huge wheels, and that wheels could return to being the same size. The end of the high-wheeler had come.

The Safety Bicycle >

The safety bicycle has a diamond frame and two equal-sized wheels, both of them far smaller than the front one of a high-wheeler (see pages 28–9). It is the basic pattern that anyone would recognize as a bike today. The word 'safety' came from the fact that its rider could sit astride the bicycle with both feet touching the floor. Also, the ride, feel and control were far better than on a precarious high-wheeler.

The safety bicycle first saw the light of day in 1876, when its rear wheel was powered by treadles. The inventor was Harry John Lawson, but his bike did not catch on because it was heavy and relatively complicated.

A number of variations were tried without success, until John Kemp Starley brought out his chain-driven version – the Rover. After a short period of development, the Rover proved such a hit that high-wheelers were largely abandoned.

BELOW An open-framed safety bicycle with a sprung leather saddle. Safety cycles saw the introduction of lever- and rod-operated brakes [1]. This model has rudimentary suspension on the forks, and also foot rests, so the rider could take his or her feet off the pedals on long, downhill sections [2]. An early roller chain drives a single fixed gear [3].

High-wheelers were the province of agile young men or very daring young women, but almost anyone could ride a safety bicycle. Starley's Rover sold in the thousands and, although the bike was soon copied, his company prospered and became known as the Rover Cycle Company.

The term 'diamond frame' comes from the safety bike's frame comprising two triangles. One is formed by the bike's top tube, seat tube and down tube. The other triangle is created by the bike's seat stays, chain stays and seat tube.

The other thing that made the safety bicycle popular was its use of the pneumatic tyre. The original smaller-wheeled machines had solid tyres and gave a harsher ride than the high-wheeler. The arrival of air-filled tyres that absorbed a lot of road shock and gripped the road better removed the last of the drawbacks associated with the smaller wheel.

John Kemp Starley >

Born in Walthamstow, Starley moved to Coventry aged 18 in 1872, to work for his uncle, James Starley, a manufacturer of one of the best high-wheelers, the Ariel. In 1877, along with local enthusiast, William Sutton, John Kemp Starley founded his own company with the aim of developing a safer alternative to high-wheeled bikes. Their design was called the Rover.

With a low centre of gravity, the Rover's rider sat between the bike's wheels and could place his or her feet on the floor while still astride the bike when stationary. The smaller, equal-sized wheels also made braking much easier. High-wheelers were the province of agile young men or very daring young women, but almost anyone could ride a safety bicycle.

Starley's Rover sold in the thousands and although the bike was soon copied, his company prospered and became known as the Rover Cycle Company. When Starley died suddenly in 1901, he was succeeded as managing director by Harry Smyth, who slowly converted the business to making motorbikes then Rover cars. Starley is so important in the history of Coventry that there is a statue of him there today.

BELOW A diamond-frame safety bicycle from the late 1870s, with the frame comprising two triangles. The shape of the modern bike is already evident in this early model.

ABOVE A poster advertising Starley's Rover, which shows that this bike could be ridden just as easily by women as men.

ABOVE John Kemp Starley (1854–1901) riding a safety bicycle. His invention was much easier to ride than the ordinary bikes that preceded it. The safety bicycle has donated design DNA to almost every other bike that came after it.

The Story of the Pneumatic Tyre >

History is littered with people who were in the right place at the right time for their ideas to catch on. Scotsman R W Thomson took out a patent in 1845 for a 'hollow rubber tube filled with air'. It was designed to make wheels roll better and could have been used on early bikes, but Thomson was unable to secure a regular supply of the thin rubber he needed to develop the idea to the point of mass production. Inventor John Boyd Dunlop knew nothing of Thomson's patent when he invented, or re-invented, the pneumatic tyre 43 years later.

Dunlop, who was Scottish but lived in Northern Ireland, found out about Thomson's work only when he sought to patent his invention – an inflated tube made from sheet rubber – which he designed for his son's tricycle. He originally made it because he thought it might prevent the headaches his son suffered when riding on rough roads. While experimenting, Dunlop discovered that pneumatic tyres rolled much farther than solid ones for the same input of leg power.

He fitted his new invention to a couple of bikes, one of which was bought by the captain of the Belfast Cruisers Cycling Club, Willie Hulme, who won every race he rode on Dunlop tyres in 1889. Commercial production of Dunlop tyres began in 1890, both for bikes and for other vehicles, and the name became soon synonymous with all kinds of tyres.

The first Dunlop tyre – a one-piece inflated tube – stayed in place by being stretched on to a wheel. Dunlop added an inner canvas layer to the rubber tube in later models, which helped to prevent punctures. In 1891 Frenchman Édouard Michelin took the idea further and invented a tyre comprising a rubber outer held on to the wheel rim with clamps, which had a separate inflatable rubber inner tube inside.

Race tyres >
Later, tyres were designed specifically for racing, called tubulars. Tubulars consist of a very light rubber inner tube wrapped and sewn into a textile casing. A canvas band covers the stitching and a rubber tread is glued around the outside circumference of the tyre. The tyre is then glued on to the wheel rim with the tread facing out. Tubulars made in this ways are still used for racing and many believe they are still the best kind of racing tyre.

Clincher tyres >
A variation of Michelin's first design, clincher tyres, came on the scene around 1930. They stayed on the wheel rim when inflated not thanks to glue but by using a continuous bead on either side of the tyre that interlocked with flanges inside the edges of the wheel rim. In the 1980s, lightweight clinchers were developed that came close to – some say surpassed – the performance and feel of tubulars.

FAR LEFT Inventor John Boyd Dunlop, with a bicycle fitted with his pneumatic tyres.

LEFT Tubular tyres were perfect for the days when competitors were not allowed support vehicles and had to carry their own spares. A tubular tyre could be folded up and fastened under the saddle, or simply looped around the rider's shoulders.

Ultimate tyres >

Lightweight tubulars (1)

These are used for racing on velodromes or in time trials, where every second counts. The light cotton or even silk casings are covered with the thinnest, most marginal tread. Silk is chosen because it is a thinner thread than cotton, so there are more threads per centimetre of casing, which means the tyre is more supple and rolls along better. Lightweight tubulars, with their strong fibres and high thread counts, can withstand very high pressures, as much as 15 bars (220psi).

Ice tyres (2)

While mountain bike tyres have studded tread patterns, ice tyres also have metal spikes in the tread to give more grip.

Self-inflating tyres

Tyres have been developed that sense if air has leaked out, which it does over time, and re-inflate automatically as the bike is being ridden.

Helium-filled tubulars

When a tubular is inflated to its maximum it contains a lot of air. Helium is lighter than air, thus a helium-inflated tubular is lighter than an air-filled one. Helium has been used to inflate tyres for a number of successful world record attempts.

1

2

LEFT Roger Rivière setting a new world hour record on Milan's Vigorelli track in 1958. His bike was designed to be as light as possible; his tubular tyres were even filled with helium because it is lighter than air.

The Safety Bicycle Explained >

It is worthwhile looking in detail at the safety bicycle's design because it is the blueprint for the modern bike. Although there have been notable deviations from it, the safety bicycle's double-diamond frame design is seen in bikes though the years, and persists in the majority of bikes today. For the background to its invention, see pages 24–5.

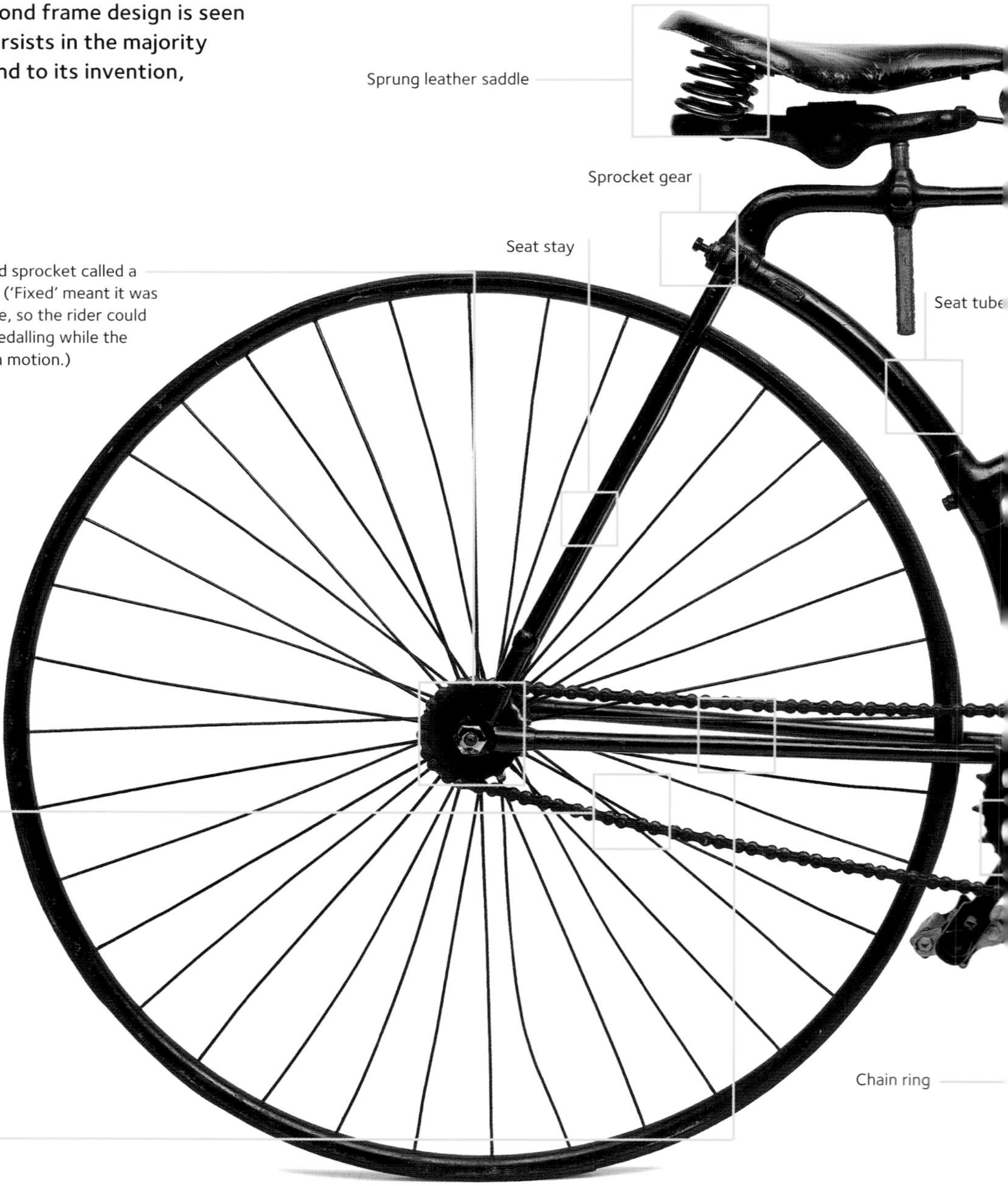

Sprung leather saddle

Sprocket gear

Seat stay

Seat tube

Single fixed sprocket called a fixed gear. ('Fixed' meant it was direct drive, so the rider could not stop pedalling while the bike was in motion.)

Chain

Chain ring

Chain stay

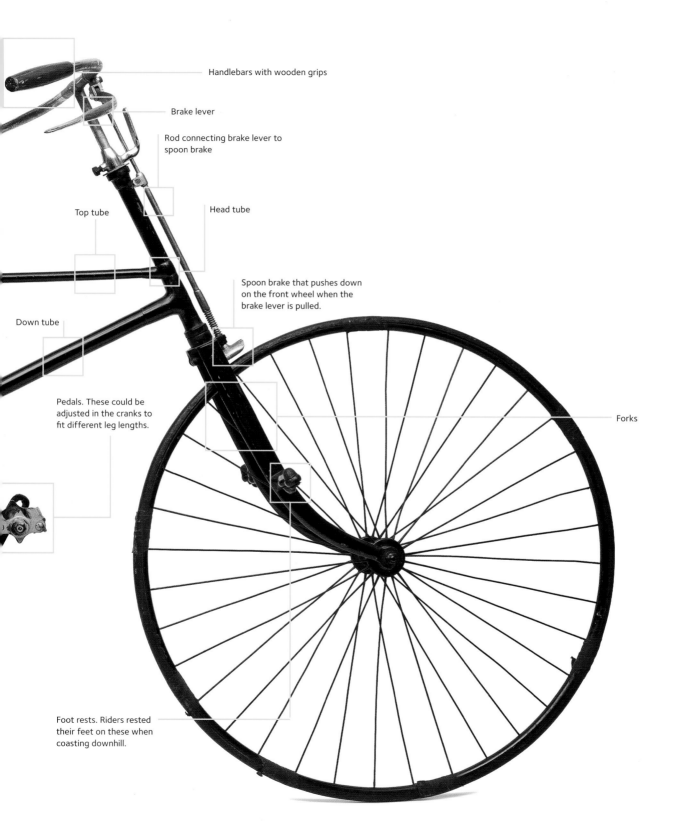

Handlebars with wooden grips

Brake lever

Rod connecting brake lever to spoon brake

Top tube

Head tube

Down tube

Spoon brake that pushes down on the front wheel when the brake lever is pulled.

Pedals. These could be adjusted in the cranks to fit different leg lengths.

Forks

Foot rests. Riders rested their feet on these when coasting downhill.

The Cycling Movement and Society >

Although they were an affordable form of transport that doubled as a new leisure activity, bicycles were not just for the working class. With the safety bicycle in mass production all over Europe and, later, in America, a cycling boom started that all sorts of people took part in. Cycling was also an adventure that members of both sexes could be involved in together.

Cycling clubs sprung up and their organized Sunday rides out into the countryside meant that remote inns and cafés that might previously have been struggling suddenly had extra and hungry customers to satisfy. A lot of the new clubs were truly egalitarian, with members from all walks of life mixing and sharing views. Some even had a political purpose.

The Clarion was a socialist newspaper that was started in Manchester in 1891 and was distributed by young boys riding bicycles. As well as containing political articles, it also encouraged working people to exercise in their leisure time. The Clarion Cycling Club formed in 1894 and had sections in Birmingham, Stoke, Liverpool, Bradford and Barnsley.

ABOVE This advertisement from the Dunlop archives, featured in *The Clarion Cyclist*, shows the company's support of the National Clarion Cycling Club, whose newspaper is full of advice for its members on how they can get more from their cycling.

RIGHT Members of the Clarion Cycling Club in November 1936. Every Easter to this day, Clarion clubs from all over the UK meet in order to ride together.

Clarion cycling is alive today, with clubs all over Britain but, where their motto used to be 'fellowship is life', today the club describes itself as the 'On-Bike Social Networking Club'. Similar national networks spread around Europe and other countries around the world under the umbrella of the Confédération Sportive Internationale Travailliste et Amateur (CSIT), which was set up in Ghent, Belgium in 1913. The CSIT now has 35 member organizations in 29 countries. In France, the Féderation Sportive et Gymnique du Travail (FSGT) was set up in 1934 and is affiliated to the CSIT.

Cycling helped push society forwards in other ways, too. Cyclists lobbied for road improvements, and better roads were essential to the development of the motorcar. Some cyclists and manufacturers played a part in developing other forms of transport. The American fathers of flight, the Wright brothers, were bicycle manufacturers before they developed the first flying machine in 1903. Former cyclist Henri Farman was the first to fly 10km (6 miles) in a figure of eight, doing so in 1908 in a Voisin biplane at Issy les Molyneaux near Paris, where by coincidence the Tour de France has its headquarters.

Cycling and the emancipation of women >

At first, because of the heavy long skirts they wore, women could only ride tricycles. Then, in America, women started cycling in a type of legging known as bloomers. Bloomers are voluminous knee-length trousers, but they made riding two-wheeled bikes possible. The fashion caught on in Europe, so for the first time men and women could ride bicycles together and some women started to go on cycling adventures of their own.

Annie Kopchovsky from Boston, Massachusetts was an early cycling pioneer. Between June 1894 and September 1895 she cycled around the world, although quite a bit of her journey was done on boats and trains. She rode her Sterling brand bike between places where the trains did not go and had an amazing adventure. It was one of the things that changed society's view of women – and also influenced how women saw themselves.

Shortly before her death in 1906, the American civil rights activist and feminist Susan B Anthony said: 'Bicycling has done more to emancipate women than anything else in the world. I stand and rejoice every time I see a woman on a wheel. It gives women a feeling of freedom and self-reliance.'

BELOW Bicycles gave women the freedom to travel independently, to dress differently and, in this daring woman's case, to acquire new skills.

BELOW Cycling influenced fashion in the early 20th century, as this trendy French magazine shows.

BELOW Clarion riders shouted the word 'Boots' to approaching cyclists. If they then replied 'Spurs', that identified the cyclists as fellow Clarion members. The Clarion's biannual members' magazine is called *Boot & Spurs*.

Suits You Sir! >

Cycling affected women's fashion, but cycle clothing in general has changed immeasurably over the years. The clothes racers now wear are at the cutting edge of design; they are innovative and made from advanced textiles. Racing has driven clothing development as much as it has the development of bikes, but it has had a positive effect on developing clothing for everyday cycling too.

Plus fours and alpaca jackets >

The first cyclists wore plus fours: knee-length trousers combined with long socks. For their upper body they had cotton shirts and tweed jackets. Cotton replaced tweed for summer cycling for both jackets and trousers. The first racers wore similar outfits only adding a jacket on colder days.

Shorts and short-sleeved vests were first introduced for track racing. Cotton was soon replaced by wool because it stretches, allowing tighter fitting shorts and tops to reduce aerodynamic drag. More importantly, wool garments tend not to chafe the skin. Wool also has better insulating properties than cotton and wicks sweat away from the body, reducing the amount of moisture that gets through to the skin.

Early road racers wore wool shorts, vests and light alpaca jackets with pockets to store food in. Their races were incredibly long: Paris–Brest–Paris was 1,200km (746 miles), in one go, requiring riders to ride through the night. If it was cold, racing cyclists wore wool leggings under their shorts.

ABOVE Champion bicycle racer Montague Holbein wears a flat cap, knickerbockers and a stiff-winged collar, typical attire for a well-turned-out cyclist on an everyday ride in the late 19th century.

RIGHT Early continental racers such as Maurice Garin, pictured here in the magazine *Le Petit Journal*, wore light wool and cotton clothing. Garin won the first ever Tour de France in 1903.

Jersey pockets and chamois inserts >

As road racing developed during the early 20th century, competitors abandoned jackets, preferring to carry food in cycling-specific wool jerseys that had pockets stitched to the chest and to the lumbar region of the back. These jerseys were short- or long-sleeved. Races were so long that riders often wore several layers for early morning starts, removing them as the day heated up. However, the rules stated that competitors had to carry any items removed with them, which caused several famous arguments between racers and officials.

Shorts were still made of wool but had chamois leather inserts to protect the area where cyclists sat on their saddles. Chamois leather is soft, resists creasing and absorbs moisture, so it helped reduce saddle sores – a problem that has been almost eradicated by the synthetic inserts used today.

LEFT By the middle of the 20th century, road racers, like these at the start of the 1936 Paris–Brest–Paris race, wore clothes that looked similar to racing clothes worn today. However, the main fabric used then was wool, not Lycra.

Slippery silk >

Track racers do not encounter the wide range of weather conditions road racers do. This held true for outdoor as well as indoor tracks because if a banked track is at all wet the surface is too slippery to race on. Clothing for track racing therefore focused on helping riders be as aerodynamic as possible, so they soon began wearing silk jerseys, which was thought to slip through the air easier than wool.

RIGHT These track racers competing in the 1967 London Six-Day Cycle Race, held inside Earl's Court, wore jerseys made from silk.

Dressed for Speed >

There was a revolution in cycling in 1984, when an important notion was proved beyond doubt – that air resistance increases exponentially as cycling speed increases until it becomes the biggest factor that a cyclist has to overcome in order to go faster. The man who realized this was the great Italian racer Francesco Moser. He had won a lot of big races and was approaching the end of his career when he broke the world hour record that had been set by the best racer the world has ever seen, the Belgian Eddy Merckx. It was a record that many regarded as unbeatable. To do it Moser not only adopted the latest training methods but also used the latest aerodynamic bike and clothing – kit that had been developed by scientists and engineers in a wind tunnel.

Clothing for cycle racing is ideally as close fitting as current technology allows although, on occasions, cycling authorities have legislated against this. As speeds increased, the drag effect of billowing jackets and trousers became more significant. Trainers and equipment developers turned more and more to aerodynamics in pursuit of victory.

Advent of the skinsuit >

The arrival of the man-made fibre Lycra, which was used to make lightweight materials capable of stretching with the body, was nothing short of revolutionary in the world of cycling. Initially, it was adopted for comfortable, crease-free and easy-to-wash shorts, but the big aerodynamic step forward that this wonder material subsequently allowed was the development of the

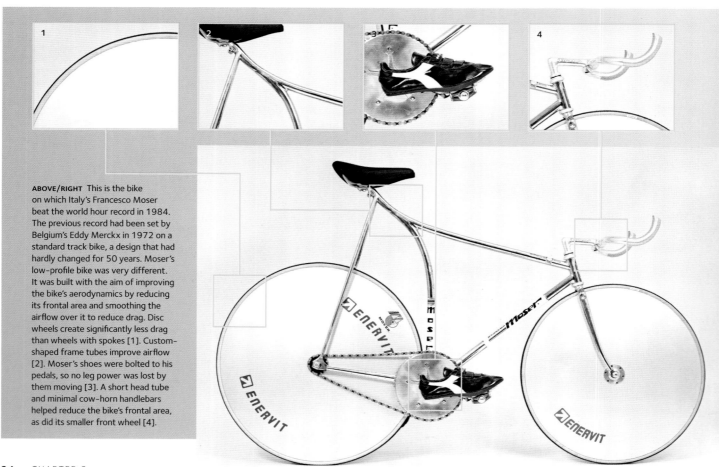

ABOVE/RIGHT This is the bike on which Italy's Francesco Moser beat the world hour record in 1984. The previous record had been set by Belgium's Eddy Merckx in 1972 on a standard track bike, a design that had hardly changed for 50 years. Moser's low-profile bike was very different. It was built with the aim of improving the bike's aerodynamics by reducing its frontal area and smoothing the airflow over it to reduce drag. Disc wheels create significantly less drag than wheels with spokes [1]. Custom-shaped frame tubes improve airflow [2]. Moser's shoes were bolted to his pedals, so no leg power was lost by them moving [3]. A short head tube and minimal cow-horn handlebars helped reduce the bike's frontal area, as did its smaller front wheel [4].

one-piece, figure-hugging garment known as the skinsuit. It was the brainchild of Swiss clothing manufacturer Toni Maier-Moussa. Combining racing top and shorts into one garment, the skinsuit removed flapping material and smoothed out creases, and was significantly more aerodynamic than a two-piece outfit. Huge amounts of research have been invested into materials and skinsuit design since Maier-Moussa invented it.

Today, a cutting-edge skinsuit might be made with a wide variety of surface textures in different places on the suit, to optimize airflow over specific parts of the body. Even seam size and placement have been researched in wind tunnels. Suits are also tailored to the riding position, making them difficult to put on and uncomfortable to stand up in (they are often shaped for the riding position). The materials used are sometimes so aerodynamic they have a lower drag than bare skin, consequently modern suits generally extend down the arms. Some suits even incorporate golf ball-like dimpled bands at the lower edge of the shorts to optimize airflow over a part of the body that is constantly changing orientation (as a spinning golf ball does).

BELOW When Francesco Moser broke Eddy Merckx's world hour record in 1984, he was clad from head to foot in a custom-made Lycra suit, designed to make him as aerodynamic as possible. Having good aerodynamics is key to breaking any such records.

BELOW British Cycling's 'marginal gains' philosophy (see page 234) is intrinsic to its international competitive effort, from the time trials of world and Olympic champion Sir Bradley Wiggins to the races of track riders such as double Olympic gold medallist Laura Trott, shown here.

BOTTOM Spain's Guillermo Timoner was best known for racing around a velodrome in the slipstream of a powerful pacing motor bike. The special clothing design that he devised – a jersey with a woollen front and silk back – helped him to be dragged a little faster.

Marginal gains 1960s style >

Looking for every advantage, no matter how slight, is not a recent mind-set in cycling. Top racers have always had this approach, as can be seen in this example from the 1960s.

Guillermo Timoner from Spain was a motor-paced racer (there'll be more about motor pacing on pages 76-77), and he raced in a jersey with a silk back and a woollen front. In motor pacing, cyclists race each other around steeply banked tracks in the slipstream of specially adapted motorbikes. Timoner's thinking was that silk smoothed out air flowing over his back, but the rougher wool created eddies of air in front of him, which helped him be dragged a little faster in the slipstream of his pacing bike.

Are You Sitting Comfortably? >

Whereas suspension systems on modern bikes are aimed at performance first and comfort second, those on early bikes were purely to make cycling a more pleasurable experience. The first metal boneshakers (see pages 12–13) had saddles mounted on a long thin plate, which acted like a leaf spring, and then pneumatic tyres helped absorb some lumps and bumps. Fortunately, it was not long before inventors came up with yet other ways of soaking up the punishment.

The McGliney patent >

On 22 December 1891 an American, C E McGliney, filed a patent for a full-suspension bicycle. The pneumatic tyre had only just got into production, and the drawings submitted of the McGliney bike seem to show solid rubber tyres. The design filled a need, although it is not known how many McGliney bikes were produced, or if it worked effectively.

McGliney's design was basically sound. The forks articulate close to the front-wheel hub around a spring-loaded pivot. The integral chain-stay and bottom bracket unit is also spring-loaded and pivots independently of the main frame triangle. Overall, the engineering of the McGliney bike design has some of the underlying DNA of a modern full-suspension bike.

Fig.1.

(No Model.)

No. 465,599.

C. E. McGLINCHEY.

VELOCIPEDE.

Patented Dec. 22, 1891.

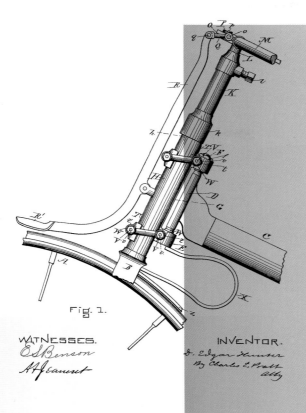

Springs and pivots >

Further suspension patents were filed throughout the 1890s. Hugo Auguste Becker designed a full-suspension bike in which the articulation of the front forks and rear triangles stretched rather than compressed springs.

Other bike designs had just rear suspension, which is quite logical on a bike with an upright riding position, as bikes were then, because shocks are transferred directly up through a cyclist's spine when riding like that.

A front-suspension bike, patented by another American, D E Hunter, in 1889 used a U-shaped metal spring attached to a rod. This, in turn, was attached by pivots to the bike's head tube, so the forks, and therefore the front wheel, moved up and down independently of the head tube.

The first telescopic forks >

Modern suspension forks have two telescopic fork legs, each comprising a slider with the front wheel attached to it and a separate stanchion. The slider moves up and down the stanchions, compressing springs inside each fork leg. Each stanchion is attached to a solid fork crown, which, in turn, is attached to the handlebars. This and variations of it are the best way current technology has developed to absorb shocks to the front of a bike. It is surprising then that there was a design for telescopic forks submitted in 1922 to the US Patents Office by George W Sage, which was similar in principle to how suspension forks work today.

These 1922 forks had two stanchions attached to a fork crown, with the sliders moving up and down inside the stanchions, compressing a spring. However, unlike modern front suspension systems, the forks had no damping system to control the spring's action. Damping is the key to why modern suspension systems work so well, because an un-damped spring adds to the difficulty of controlling a bike.

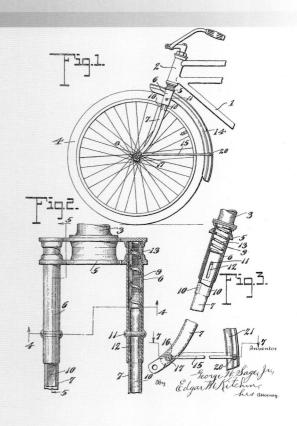

Lighting the Way >

Lights are needed when cycling at night, both to see the way, and to be seen. Soon after the modern bike was invented in the late 19th century (see pages 24–5), the use of bike lights was enshrined in law. Since then, lights have varied in effectiveness, although some of the earliest versions were very bright, even if they were powered by some pretty startling chemistry.

The very first bike lights were the oil lamps that were made for high-wheeled bikes (see pages 16–17). Then, in 1889, acetylene gas lamps were developed. They were known as carbide lamps because the acetylene gas, which the rider ignited with a match or taper, was produced by combining calcium carbide with water. That produced a volatile reaction. Once the acetylene was lit and the lamp window closed, the contents fizzed and popped inside the lamp as the cyclist rode along the road. The light produced by these lamps was very bright and the beam very strong.

Electric bike lights were introduced in 1898; they used an incandescent bulb powered by lead-acid batteries or a dynamo.

Improvements >
Both carbide and electric bike lights improved through the years. Lead acid was replaced by dry-cell then alkali batteries, which were more convenient but the light they gave lagged behind acetylene gas for a long time.

ABOVE Gas lamps were heavy and needed substantial brackets to mount them on a bike, but the beam of light produced was strong and bright.

ABOVE Early dynamo lights were powered by a small wheel that turned on the tyre as the rider pedalled. The problem was that when the bike stopped, the light went out.

Dynamos also became more dependable, and the problem of the lights going out when a bike was stationary was solved, too, as the lights were fitted back-up batteries or capacitors. Throughout the 20th century, German law drove improvements to dynamos. Until 2013, it was illegal to ride a bike in Germany without a dynamo lighting system on it, even in daytime, unless it was a racing bike weighing less than 11kg (24lb). The law forced German manufacturers to create lighter, more dependable and reduced-friction dynamos.

RIGHT An early 20th-century acetylene gas lamp. There's a window in the top, so the rider could check that it was lit, as well as side windows – green on the right, red on the left – to ensure that the cyclist could be seen from the side.

FAR RIGHT A page from a 1920s Lucas catalogue showing two kinds of battery lamps, both of which are mounted on sprung brackets to protect them from potentially damaging vibrations.

Modern lights >

Halogen bulbs were used in bike lights from the 1980s and were a big step in improving bike lights. They needed a lot more power than standard bulbs, so rechargeable batteries became widely used, as well as even more efficient dynamos. The beam from a halogen bulb is so bright that cyclists had improved visibility, both for them to see and others to spot them.

Most bike lights now have Light Emitting Diodes (LEDs) as their light source, LED lights having come into use for bicycle bike lights during the 1990s. These are very reliable, bright and require less power than any equivalent bulb. They are used in front white lights and rear red lights, as required by law in most countries of the world. LEDs are often multi-functional and can be used in constant flashing mode. They are also very light, meaning little weight penalty for a cyclist carrying two or more front and rear lights. Some riders attach LEDs to their clothing and to their helmets, too.

RIGHT A selection of lights available for bikes today. The simplest merely help a rider to be seen, while the very powerful are capable of casting a wide, strong beam that allows a rider to see well enough to ride off-road in pitch-black conditions.

Building the Brand >

The popularity of cycling spawned hundreds of bike manufacturers, many of which are now no more than names on museum pieces. But some survived the cycle industry's boom-and-bust history, and are still enjoying success today. This is the story of three of those iconic brands.

France: Cycles Peugeot >

Peugeot was originally a steel manufacturer that made knives and forks. Seeing the rise of the bicycle as a possible income stream, the company created Cycles Peugeot in 1882, using its existing lion's head trademark as the company logo.

It was one of the first manufacturers to sponsor professional racers and in 1905 Louis Trousselier won the third Tour de France while riding a Peugeot bike. The company soon formed a cycling team and became the most successful bike supplier in Tour de France history, achieving some ten victories.

As the sport of professional cycling grew so did the cost of sponsoring a team. Although other bike manufacturers became secondary sponsors by supporting teams financed by companies from outside cycling, Peugeot carried on as a headline team sponsor until 1986. Their white kit with a black chequered band around the middle was one of the most iconic in the sport. Even after it stopped being a main sponsor, Peugeot carried on supporting pro road racing teams and is a major bike producer to this day.

ABOVE The great Eddy Merckx, in Peugeot black and white, racing alongside Italy's Gianni Motta in the 1966 Milan–San Remo race.

ABOVE The Peugeot team line-up at the start of the 1966 Tour de France. The rider not in the team's black and white kit is Britain's Tom Simpson. He is wearing the world champion's rainbow jersey.

United Kingdom: Raleigh >

In 1887, a prosperous Nottingham businessman called Frank Bowden was advised by his doctor to take up cycling. He bought a bike from a manufacturer on Raleigh Street in the city, and was so impressed by it that he bought the business. He moved premises, changed the company name to Raleigh, and within six years it was one of the biggest bicycle manufacturers in the world. It also earned Bowden a knighthood.

After Sir Frank Bowden's death in 1921, Raleigh was run by his son Harold Bowden, who kept on expanding the company. By the time the Second World War started in 1939, Raleigh had spread across a 3-hectare (7½-acre) site, and was one of Nottingham's largest employers.

Interest in cycling as transport dwindled after the war, but Raleigh maintained its success by changing some of its production to lightweight race bikes. Part of its success was due to Raleigh's sponsorship of the first racing cyclist to become a household name in Britain – Reg Harris – and its advertising campaign with the catch phrase 'Reg Rides a Raleigh'.

Inevitably, despite peaks of production in brand leaders such as the Chopper (see pages 112–13), the business shrank and Raleigh went through various stages of restructuring until it no longer occupied its giant site. The brand is still alive and its profile is growing again. The company headquarters is still in Nottingham although bikes are no longer made there.

RIGHT Reg Harris played a key role in Raleigh's publicity campaigns in the 1940s and '50s. Harris was the first cyclist to become a household name in Great Britain.

America: Schwinn >

In 1895, two German immigrants to America – Adolph Arnold and Ignaz Schwinn – founded Arnold, Schwinn & Company, a bike manufacturer that later became the Schwinn Bicycle Company. Its ambition was to make money from the sudden craze for cycling, which had exploded in the United States.

The craze was short-lived because, by the time cars were being mass-produced, Americans preferred them to bicycles. By 1905, cycle sales were 25 per cent of what they had been at the height of the bike boom just five years earlier. Many manufacturers went out of business. Schwinn, however, realized that the only way to survive was to grow and take a greater market share and he successfully steered the company through a difficult transition period.

Schwinn was dealt another blow by the stock market crash of the 1920s, but by selling off his motorcycle division – which once rivalled Harley-Davidson – and concentrating on his core business, Schwinn grew the bike business again, making a type of bike unique to America called the cruiser (see pages 100–1).

Schwinn began producing lightweight race bikes in the 1960s, at one time supplying them to America's first professional cycling team, 7-Eleven, but the company dropped out before the team made its Tour de France debut in 1986. Schwinn cruiser bikes adapted by off-road riding enthusiasts and called klunkers were the first mountain bikes.

Unfortunately, Schwinn suffered from trade union problems and the company was badly affected by lower-cost products from the East Asia. Despite a pioneering role in American race bikes and mountain bikes, the company was not as innovative as new manufacturers such as Trek, Specialized and Cannondale. Thus, in 2001, Schwinn and its assets were bought by Pacific Cycle. The Schwinn name is now put on bikes made in China.

The Bamboo Bike >

Modern bamboo bikes are perfect machines for 21st-century personal transport. The material they are made from is sustainable, so using a bamboo bike is about as eco-friendly a way to go cycling as possible. And bamboo works well in bicycle construction as it is strong, stiff and absorbs vibrations. It is also easier to repair than other materials. The first bamboo bikes were made way back in 1892 and are once again enjoying attention from the cycling community as, since the beginning of the 21st century, bamboo has had a resurgence.

The Bamboo Bicycle Company was set up in 1892, and in a brilliant piece of marketing a number of bikes were sent to titled families, so that the company could use any positive feedback in its advertising. Lord Edward Spencer Churchill, a cousin of the Second World War British prime minister Sir Winston Churchill, wrote: 'I am pleased to state that I rode the Bamboo cycle you supplied me with last December, for about 1,500 miles and am very pleased with it. It is quite the best for hill climbing I have ever had.' Functioning in much the same way as a celebrity endorsement could do today, it boosted the popularity of the bamboo bike.

RIGHT/BELOW An early bamboo bike. Its well-sprung saddle can be moved forwards and backwards on the seat post, as well as up and down, to help provide a comfortable position [1]. Bamboo main tubes are joined through metal joints [2]. The bike even has wooden mudguards [3].

Eco-friendly bikes >

Efforts have been made over the last few years to find alternatives to expensive and in some cases non-sustainable materials used to make bikes. One of the more recent and most interesting – because its inventor believes one day it could retail for as little as US$20 – is the cardboard bike.

Its inventor and Israeli engineer Izhar Gafni described the construction methods required to make a bike almost entirely out of cardboard: 'It is very strong once it is folded. In the end the solution is like Origami, folding it triples cardboard's strength.' The bikes are finished with a coating that protects them from the effects of weather.

Bamboo bikes were lighter than 19th-century steel bikes, although the use of bamboo was limited to the frame tubes. These were joined to each other through metal lugs, in a similar way to how the first commercially available aluminium frames were produced 80 years later. All bamboo bikes had wooden wheel rims – as all bikes did then.

The Bamboo Bicycle Company did not last for long. Early bamboo bikes were not very rigid, so they did not transfer power efficiently and steel was seen as a better, more modern material. However, the use of bamboo in bike design has made a comeback since the the turn of the century and a company called Calfee Design now manufactures top-of-the-range road-race bikes, mountain bikes and tandems, as well as some Fixie models, all made from bamboo. Meanwhile, the Bamboo Bicycle Club (www.bamboobicycleclub.org) can help you to build your own bamboo bike.

ABOVE/RIGHT A modern bamboo bike, with details showing the exquisite way in which the main tubes are joined.

Gearing Up >

The arrival of gearing caused interest in cycling to explode and meant that bikes were no longer dependent on wheel size for speed. The gearing of the first safety bicycles was a single ratio. However, experiments with multiple ratios had already been carried out on high-wheeled bikes (see pages 16–17) and in 1878 a high-wheeler was produced with the first epicyclic hub gear system. It was not long before multi-geared systems were produced for the safety bicycle (see pages 24–5) and the models that followed it. Four inventions allowed the development of multi-ratio gears for bikes.

The freewheel >

Invented in 1869, the freewheel allowed a cyclist to stop pedalling while his or her feet were still in contact with the pedals. A freewheel is a toothed sprocket on the rear wheel that is driven by the bike's chain. When a cyclist pedals forwards, a mechanism engages inside the freewheel body and dives the bike. When the cyclist stops pedalling, the mechanism becomes disengaged and the rear wheel continues turning as the bike coasts along. The freewheel enabled designers to think about fitting multiple gears to bikes.

The gradient gear >

This is the earliest ancestor of the modern derailleur gear. Invented by E H Hodgkinson in 1896, it was a device that lifted the chain and shifted it between three sprockets on the rear wheel. The rider pedalled backwards to derail the chain, then operated a lever that shifted the chain up or down a sprocket. Any slack in the chain was taken up by running it through a small spring-loaded wheel, rather like the two small jockey wheels on a modern derailleur gear.

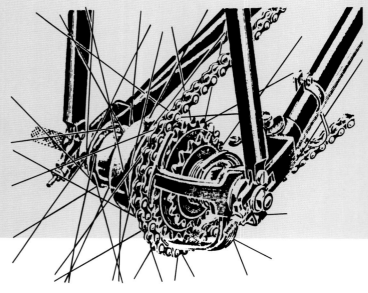

Sturmey Archer three-speed gears >

Sturmey Archer started producing three-speed hub gears in 1902 and still does today. The company, which was founded by Henry Sturmey and James Archer, started making a version of William Riley's epicyclic two-speed gear. The word 'epicyclic' describes a gear system where the cogs of a central shaft engage with different-sized planetary cogs – three in the case of the first Sturmey Archer hub gear but more in later versions. Sturmey's most famous product, the AW – which was first produced in 1936 and continued until 2008 – is a model of design maturity.

Hub gears require little maintenance and are very convenient, but they are also heavy. There is still plenty of demand for them and a number of other companies, including Shimano and Rohloff, make excellent hub gears.

ABOVE A bike equipped with a Sturmey Archer three-speed hub gear. Details of the hub unit and the handlebar-mounted trigger shifter are shown.

Reproduction F. Meyer - Carpentias

Paul de VIVIE dit " Vélocio "
Père du Cyclotourisme
1852-1930

Paul de Vivie >

De Vivie was a French bike importer and cycling magazine publisher who invented the double chain-set, which later made 10 and now 22 gears possible on road bikes. However, he initially used the double chain-set as a simple solution to riding uphill. De Vivie lived in Saint Étienne, at the foot of the Col de la République, and struggled to ride up it with the single gear ratio he used every day to get about town. So he added a smaller chain ring inside the standard one, and merely picked up the bike's chain before a hill and placed in on the smaller chain-ring, thereby giving him a lower gear ratio.

Later, de Vivie added another double chain-set and chain drive on the left-hand side of his bike, giving four gears. Then he adapted a gear shift system from Britain called the Proteon on a bike called the Whippet, in which back-pedalling made the chain-ring change size through a system of internal pawls and springs. Once normal pedalling was resumed, the rider had a different gear ratio.

LEFT Paul de Vivie, pioneer of derailleur gears, is shown here with one of his inventions from 1911: an early small-wheeled bike.

Chapter 3
Climbing Mountains >

The Italian Gino Bartali in the 1950 Tour de France. He won this stage but later left the race, along with the whole Italian team, after being threatened by French supporters who were still angry about Italy's role in the Second World War. Bartali won the Tour de France in 1938 and 1948 and was actually a member of the Italian resistance during the war, working against the Nazis.

Early Shifts >

By the end of the 19th century, multiple gears were being widely used by all but pro racers, who thought they were for tourists and softies and that pedalling a single ratio over challenging terrain was how 'real men' went cycling. However, the inclusion of mountains in races such as the Tour de France made the pro racers think again, although outdated 19th-century attitudes to multiple gears were preserved well into the 20th century by the organizers of the Tour de France.

Although the first bicycle races were on fairly flat terrain they covered a very long distance. When the Tour de France started, getting around the whole hexagon of France involved crossing some high places. The first Tour de France in 1903 climbed over the 1,161-m (3,800-ft) Col de la République, the same climb that spurred Paul de Vivie to invent his version of the derailleur gear (see page 45).

Early racers did have a choice of gears. Their bikes had two – or, in some cases, three – different-sized sprockets, on the same side of the rear hub, or sometimes they had sprockets on both sides. The smaller sprocket was for riding on the flat and up slight inclines. If a hill was too steep, riders stopped to undo the nuts that secured the rear wheel in place. They then placed the chain on the larger sprocket, giving them a lower gear ratio for the steep uphill section; after retightening the wheel nuts off they went again. At the top of the hill, so that they had a higher gear to ride down the other side, they had to reverse the process.

Sprocket choice and when to flip a rear wheel became part of the tactics of professional cycling. The rider who best matched his two- or three-gear ratios to the race profile had an advantage. So did physically strong riders who avoided having to stop.

This arrangement worked fine for a while but once really long and high mountain routes were included on the Tour de France route – from 1910 onwards – even the racers thought that they should be using multiple gears. Some organizers agreed but the Tour de France did not. The privateers who took part and were classified as tourist routiers (see page 88) could use them but the pro riders in the big teams, and later those selected to take part by their national federations, were not allowed multiple gears until 1937.

RIGHT Léon Scieur replaces his rear wheel after mending a puncture in the 1921 Tour de France. Riders at this time had to undo and tighten their wheels using wing nuts, which was sometimes difficult to do, especially in cold weather.

Sprocket choice and when to flip a rear wheel became part of the tactics of pro cycling. The rider who best matched his two- or three-gear ratios to the race profile had an advantage. So did physically strong riders who avoided having to stop.

The sprocket tattoo >

Tour de France rules were draconian when the competition began and remained so for many years, with race directors acting as judge and jury right into the 1970s. There were race referees – or commissaries as they are called in cycling – but their decisions were not only influenced by the race director but could also be over-ruled.

By the 1970s, most of the toughest rules had long gone but one remained. Every rider had to finish each stage with every scrap of clothing and equipment he had started out with. This included any broken bits of bike. It was worse initially: riders were not allowed to replace broken parts so they had to mend them en route. If a competitor could not and it was too far to walk to the finish, he was disqualified.

In 1921, so many spokes broke in the rear wheel of Léon Scieur of Belgium that it went totally out of shape and he could not mend it, so he had to use a spare. Yet the Tour rules still forced him to carry the original wheel, so he made a sling for it with some handlebar tape and carried it on his back for 300km (186 miles) to the end of the stage. Scieur was left with a star-shaped scar on his back from where a sprocket had dug through his skin and it remained for the rest of his life.

ABOVE The loneliness of the long-distance cyclist. Émile Angel had a mechanical problem in the 1913 Tour de France but Eugène Christophe, who was about to pass him, was forbidden to offer assistance, even though he was in the same team. Doing so was against the rules.

Racer Inventors >

Tullio Campagnolo was born in 1901 to a middle-class family in Vicenza, Italy. He grew up tinkering with inventions in his father's hardware store, and racing on his bike. He was a quite successful cyclist but, as time would tell, he was even more talented as an inventor. Campagnolo is the best known of the racers who have improved bike design by developing items to help their own performance.

On 11 November 1927 Campagnolo took part in a race in the Italian Alps. Racing on mountain roads at that time of year was hazardous. Despite the snow on the ground, by the time he reached the Croce d'Aune pass Campagnolo had built up a good lead when he suffered a rear-wheel puncture. At that time wheels were secured by wing nuts – as a rider could undo a wing nut without a spanner. But Campagnolo's hands were frozen and he could not grip the wing nuts with enough force to turn them. He was forced to abandon the race and vowed to find a better way to secure a bike's wheels in its frame.

Quick release >

By 1930, Campagnolo had invented, developed and was producing a lever-operated, quick-release mechanism that is still used to fasten bike wheels in frames today. In order to manufacture his invention, the inventor founded the Campagnolo Company, which has helped drive the development of bike parts ever since (see also pages 52–3).

Other inventions by racers >

No other cyclist has had the effect on bicycle development that Tullio Campagnolo had but several others also came up with innovations that are taken for granted today. Australian sprinter John Nicholson created a version of the clipless pedal in the early 1970s (see pages 174–5). He adapted the type of pedals that racers used, which had toe clips and straps to secure their foot to each pedal. Nicholson's design fitted a flat metal plate to each axle then bolted his shoes directly to the plates. As his design fixed the rider's shoes permanently to the bike, his one-off invention was suitable only for sprinters and other short distance track racers.

Looking for an alternative to the heavy leather saddles that were prevalent in the 1960s, Britain's Tom Simpson covered a cheap plastic one – which was light but very hard – with a thin layer of foam rubber and a layer of leather that had been cut from his wife's old handbag. Saddles are still made like that today ... although not from old handbags.

FAR LEFT A vintage Campagnolo quick-release mechanism, fastening the wheel to the frame. Its design has changed little since it was invented in the late 1920s in response to the difficulty racers faced undoing the wing nuts that secured their wheels at the time.

LEFT Tullio Campagnolo, the inventor of the quick-release mechanism that bears his name. Campagnolo also developed the derailleur gear and founded a company that remains at the forefront of bicycle component design and manufacture today.

Tullio Campagnolo grew up tinkering with inventions in his father's hardware store, and racing on his bike. Campagnolo is the best known of the racers who have improved bike design by developing items to help their own performance.

When history repeats >

The cycling historian John Pinkerton once commented: 'Think of a new idea in bicycle design and someone will have already invented it.' And he was right. Often components and ideas that are hailed as the latest technological leap and which go on to become an accepted part of the bicycle story have already been thought of many years before.

That was certainly the case for a Sheffield-based pro cyclist Wes Mason. As well as being the 1962 Commonwealth Games cycling road-race champion and winner of many other races, Mason was a skilled frame builder. When he raced for a British professional team sponsored by Carlton Cycles, he also worked for the company and in the late 1960s he delivered a project to build the lightest possible bike for Carlton.

Only two prototypes of the bike were ever made. Among other innovations, the frames had integrated seat posts and brakes that pivoted directly on the forks and seat stays. These two weight-saving devices are used by a number of bike manufacturers on their top models today and are heralded as innovations without the knowledge that Mason had already invented them 40 years earlier.

BELOW A modern carbon-fibre frame with an integrated seat post. This makes a bike lighter and stronger than a bike with a separate seat post.

LEFT Racer and inventor John Nicholson (right) is narrowly defeated by Daniel Morelon at the Munich Olympic Games in 1972. Note the absence of toe clips and straps on Nicholson's bike.

Evolution of Gears: The Early Years >

It is impossible to underplay Tullio Campagnolo's role in this part of the bicycle's story (see also page 50). From the early 1930s to the 1970s, Campagnolo drove development of the derailleur gears as well as many other components. Other companies developed ideas of their own – and some were successful – but Campagnolo was the gold standard for a long time. Since the 1970s the company has experienced mixed fortunes but it still occupies a revered position in cycling. Many cyclists still refuse to ride with anything else.

Cambio Corsa >
This was Campagnolo's evolution of the gradient gear. It was a rickety contraption by today's slick standards, but it worked.

The bike had two rods that reached halfway up the right-hand seat stay. When the rider wanted to change gear, he reached down and turned the handle on one of the rods, releasing the rear wheel, which was held in notched dropouts. The rider then operated the second rod and, by back-pedalling, derailed the chain to a larger or smaller sprocket, depending on which way the rod was turned. At the same time, the wheel moved forward or back in the dropouts, to preserve chain tension. The rider then secured the wheel using the first rod and resumed pedalling. It sounds awful and messy but for racers at least it was revolutionary, as it allowed them to change gear without getting off their bikes.

Development of the Osgear >
Campagnolo's Cambio Corsa had a number of rivals, notably from another former racer, Oscar Egg – the Osgear. When derailleur gears were finally permitted for all competitors in the 1937 Tour de France, the Osgear was the only system allowed in the race. (In those days, all Tour de France bikes were supplied by the race organization.) In 1937 Roger Lapebie won the race using an Osgear derailleur.

The Osgear maintained chain tension in a similar way to the gradient gear, by using a lever and cable to adjust a small wheel through which the chain ran. The lever pushed the wheel down to take up any slack when the chain was shifted to a smaller sprocket, while reversing the procedure facilitated shifts to larger sprockets. The shift was made by a two-pronged fork operated by a shift lever on the bikes frame and connected by a cable. Crucially, because the small wheel took up chain slack, the wheel could remain in the same place in the dropouts.

LEFT Close-up of Campagnolo's Cambio Corsa derailleur gear, showing the operating levers and choice of four sprockets. The upper lever releases the rear wheel, while the lower shifts the chain.

BELOW Campagnolo's Nuovo Record chainset and derailleurs. This was the best equipment that money could buy in the 1970s.

Campagnolo's Cambio Corsa had a number of rivals, notably from another former racer, Oscar Egg – the Osgear. When derailleur gears were finally permitted for all competitors in the 1937 Tour de France, the Osgear was the only system allowed in the race.

Tullio's answer >

Campagnolo worked on the Cambio Corsa whenever possible through the Second World War, and, once bike racing resumed, the best prewar Italian racer, Gino Bartali, won the 1948 Tour de France using a newer, more sophisticated version.

This system was two chain-rings and four sprockets, with front shifts made by a lever that was part of the front gear shift mechanism. It pivoted in the centre and when the knob on the end of the lever was turned inwards, towards the frame, the chain shifted to the big ring. When the lever was moved the other way, the chain shifted to the small ring. Two shift levers were mounted on the bike's down tube, and these operated the sprocket shifts and the chain tensioner through control cables.

ABOVE A Gino Bartali bike from 1949, showing the Campagnolo Gran Sport rear mechanism (see page 54).

LEFT Close-up of the Osgear's chain-tensioning arm. The bike's chain runs through a small pulley wheel attached to a spring-loaded metal arm. This device moves up and down, according to the size of sprocket that the chain is shifted to. At the same time, it exerts force on the chain to preserve tension in it.

Evolution of Gears: Bringing the Story up to Date >

Campagnolo made one more giant step forward when the company developed the first parallelogram rear mechanism – the Gran Sport – in 1951. Its parallelogram action and double jockey wheels are how all rear mechanisms (or rear mechs for short) operate today. The Gran Sport rear mech was cable operated by down-tube shift levers, as was the Gran Sport front mech, providing a choice of eight gear ratios at first, but ten later on. Campagnolo carried on through the 1960s, 1970s and into the 1980s, refining its gear shifters and other bike parts until the zenith of the technology – its elegant Super Record group-set. Super Record was engineering raised to art but Campagnolo now had serious competition in areas of innovation.

Index gears >

Japanese company Suntour launched a system in 1969 where one click of the shifter equalled one gear change. These became known as index gears. Before this, shifters worked on friction; making accurate gear changing a skill. It was also common with friction systems for forces acting on the chain to pull the shifter out of position while the rider pedalled, causing the rear mech to drop the chain into a higher gear. This happened in races a lot when riders were making big efforts uphill and was exactly where a sudden unplanned shift into a higher gear was the last thing they needed. The arrival of the indexing system allowed anyone to change gears accurately with a single click.

Suntour was a true pioneer. The company also developed a rear hub with an integral freewheel body, to which a cassette of sprockets was added. (This is how rear hubs work today.) However, it was another Japanese company – Shimano – which had greater success with index gears. It developed similar ideas to those of Suntour but marketed them in a way that meant they became more readily adopted.

Shimano's indexed gear shift was named Shimano Index System (SIS) and in 1989 it released a revolutionary combined shift and brake lever called Shimano Total Integration (STI). After that, cyclists did not even have to take their hands off their handlebars to change gear.

Campagnolo played catchup through this period of development and eventually produced products, such as its ErgoPower brake-and-shift levers, to rival those of Shimano. Since 1988, but especially after the company received outside capital investment in 2008, another manufacturer – SRAM –

has shared a much bigger market with Shimano and Campagnolo. SRAM's integrated gear shift system is called DoubleTap.

Electronic gears >

The bicycle boom in the USA in the 1970s took European manufacturers by surprise, allowing Shimano and Suntour an opportunity to grow and become established in the market. After its initial breakthrough, Shimano continued to innovate faster and more effectively than the incumbent equipment giants of the day and for a while even the legendary Campagnolo, which had had little competition to this point, had to follow rather than lead. Although today both companies are on an equal technological footing with their Di2 and EPS electronic gear systems, it was French company Mavic who first introduced the concept of electronic gear-shifting to the racing world when it launched its Zap electronic rear gears in the early 1990s.

Between 1994 and 2000, I used the Zap and its later incarnation the Mektronic system extensively on road and time-trial bikes. It was most advantageous in the time-trial disciplines, where, by use of a remote button on the end of the triathlon bar, the gears could be quickly and accurately changed with a mere twitch of a finger. I utilized the system in 1994 to win the Tour de France prologue, with a record average speed that still stands.

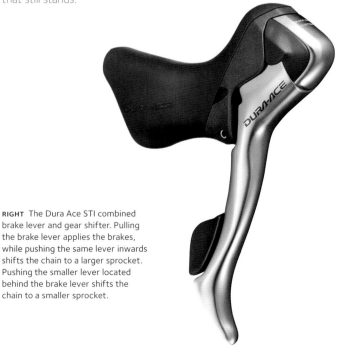

RIGHT The Dura Ace STI combined brake lever and gear shifter. Pulling the brake lever applies the brakes, while pushing the same lever inwards shifts the chain to a larger sprocket. Pushing the smaller lever located behind the brake lever shifts the chain to a smaller sprocket.

Ironically, as the potential for the system was seen and Mavic committed a huge amount of resources to take it from crude prototype to consumer-friendly product, the Zap became more complex and often failed. The pro riders lost confidence in it and the system fell from grace. Without a shop window to convince the public that Zap, and subsequently Mektronic, represented the future of gear changing, the company ceased production of the system.

Several years later, electronic gears appeared back on the market with the arrival of Shimano's Di2 and Campagnolo's EPS (Electronic Power Shift), and these systems were much better received by cyclists. Both of them, presumably due to the fact that Mavic still held patents, used battery power rather than employing rider muscles to change gear, so both had a significantly larger battery pack than Mavic's system. Unlike Mavic, both Campagnolo and Shimano opted to electrify the front mechanism, too.

Although it is highly debatable whether powered changing actually improves cycling performance, the new technology – now very reliable – captured everyone's imaginations and quickly became mainstream. Every major frame manufacturer now makes its top road frames compatible with at least one of the major manufacturers' gearing systems, supplying internal routing and battery-mounting points as standard.

The cassette >

At the same time as the method for changing gears was being developed, so was the amount of sprockets available. In the 1960s, cyclists had a maximum number of six rear sprockets to choose from. The seven-sprocket freewheel was introduced in the early 1980s, and, by the mid-1990s, sprocket choice had jumped to eight, then nine. Progress continued through the millennium to the current 11-speed cassette. To accommodate the added number of sprockets and to allow for further progress, such as being able to change gear smoothly while out of the saddle, sprockets were now wafer thin, carefully sculpted and required specially designed chains.

RIGHT Campagnolo's Chorus groupset EPS electronic brake lever and shifter. A lever behind the brake lever is moved inwards to make shifts to larger sprockets and chain-rings. The trigger on the inside of the brake lever hood is pushed down to shift the chain on to smaller sprockets or chain-rings.

LEFT The author warming up for a time trial on a bike equipped with Mavic electronic gears.

Evolution of Hub Gears >

This is a very different story to the evolution of derailleur gears (see pages 52–3), for one simple reason – Sturmey Archer's AW three-speed hub gear, introduced in 1936, worked so well that it was hard to improve on (see pages 44–5). Fichtel & Sachs made a similar gear in Germany, and together with Sturmey Archer they dominated the world market for hub gears until the early 1990s.

By 1990, Shimano had developed its hub gear system, the number of ratios had begun to increase and interest began to grow again. In 1995, Fichtel & Sachs released a 12-speed hub gear; then, three years later, another German company, Rohloff, introduced its 14-speed Speedhub. Its spread of ratios was the equivalent of having three chain-rings and nine sprockets, a set-up common on mountain bikes at the time.

The Achilles heel of hub gears is the amount of power lost through friction. Rohloff ameliorated this problem somewhat by replacing the traditional round bearings in the system with low-friction needle bearings.

Hub gear shifts are made via a handlebar-mounted trigger or twist-grip, but in 2007 Nu Vinci began manufacturing a

ABOVE Early Sturmey Archer hub gears. The small chain coming out of the right end of the rear axle attaches to the control cable and is called an indicator spindle. The chain pulls or releases a rod according to the pulled or released position of the control cable. This makes the gear shifts inside the hub.

wholly new system aimed at commuters. Their stepless shift hub brought automatic gear changing to the bicycle world. Since then SRAM, Sachs' successor in the field, introduced a hub/derailleur gear hybrid. This hybrid system fills a niche in cycling today on folding, commuter, utility and touring bikes.

RIGHT Shimano's Alfine hub gears are available with eight or eleven different ratios and are aimed at commuter cyclists.

Hub gears: pros and cons >

Pros

Hub gears are sealed from the elements, thereby requiring less maintenance than derailleur gears. Gear shifts entail a straightforward low-to-high and back down again, like driving a car. They can also be made when the bike is stationary, which is of benefit to commuters. The chain line never changes, so hub-gear chains last longer and are easier to construct, making maintaining a hub-gear system cheaper than looking after derailleur gears. Overall, hub gears are less prone to wear.

Cons

Even when riding at a steady pace, hub gears are about 8 per cent less efficient than derailleur systems at transferring the power needed. This inefficiency rises with increased power input and is the main reason hub gears are not used on race bikes. In general, hub gears are more costly than all but the most expensive derailleur systems. Hub gears are also heavier than derailleurs. In the event of a puncture, an inner tube cannot be replaced without disconnecting a hub gear's control cables, making them slightly more complex to use in the field.

ABOVE SRAM G8 hub gears are operated by a twist grip on the handlebars and provide an instant shift that rivals derailleur gears.

RIGHT/BELOW The Rolhoff Speedhub 500/14 is a masterpiece of precision engineering. It was developed by Bernard Rolhoff after his derailleur gears became clogged with sand and wouldn't work during a cycling holiday in 1994.

Tandems are still used in top-level racing. A sighted pilot steers a visually impaired stoker in most of the disciplines open to able-bodied riders. This action is from the sprint final at the Glasgow 2014 Commonwealth Games, as Australia (in front) take on Scotland.

Two Wheels or Three? >

Tricycles, or trikes as they are affectionately known, were first made at the time of the high-wheeler (see pages 16–17), often as an alternative for women and children to enjoy cycling. Tricycle riders did not sit on top of their high wheels, but between them, so the trike's speed and efficiency were not dependent on the rider's leg length. The safety bicycle (see pages 24–5) robbed tricycles of some of their utility, leaving them as an alternative for the very young and very old, although the tricycle continues to have adult enthusiasts all through the story of the bicycle. Sophisticated race trikes play an important part in para-cycling and the Paralympics today.

RIGHT/BELOW A racing tricycle from the 1970s, with insets showing details of the rear gear mechanism driving the left rear wheel only (the rear right rolls free) [1] and the two brakes required for the bike to be legal for road use, acting on the front wheel only [2].

The earliest tricycle was thought to be a hand cycle that a handicapped man made for himself in Germany in the 18th century. Leg-powered trikes were developed at the same time as early bicycles and by some of the same people. James Starley, a pioneer manufacturer of high-wheeled bikes, whose nephew John Kemp Starley pioneered the safety bicycle, was an enthusiastic trike rider and manufacturer.

Tricycles have always been used for children's first bikes, or as bikes used in businesses such as pedal-driven taxis – and some people simply prefer three wheels to two. Adult tricycles have generally followed the development of two-wheeled bikes,

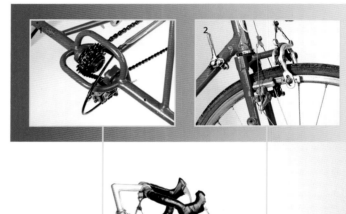

so much so that at one time it was possible to buy a trike conversion kit to turn a race bike into a race trike.

Of course, there are obvious differences: for example, trikes must have two brakes. Until disc brakes were developed, trike riders had to make individual adaptations to fit brakes to the rear wheels, so in most cases both brakes acted on the front wheel. This was achieved by combining two kinds of brakes, such as a side-pull calliper brake with a centre-pull cantilever (see pages 122–3), with each control cable going to standard brake levers.

It was not a problem that most trikes did not have rear-wheel brakes, because trike enthusiasts generally did not like rear brakes. The two rear wheels of a trike do not carry the same load as the rear wheel of a bike. Consequently, they do not have the same down-force under braking and so have a tendency to lock.

BELOW Tricycles are nearly as old as the bicycle. This early version dates from around the same time as the ordinary bicycle.

The 21st-century racing trike >

David Stone is a multi-Paralympic gold medallist, who has cerebral palsy and so races in a category that uses trikes. Paralympic trike racers participate in both massed-start and individually timed events.

Stone's trike – dubbed the Super Trike by the Great Britain team – utilizes a carbon-fibre Boardman time-trial frame with a specially modified rear section to take the dual rear wheels. For time trials, the bike is equipped with an aerodynamic handlebar set-up, exactly the same as those used by able-bodied time triallists. Similarly, standard aerodynamic deep-rimmed wheels are used to improve airflow. For mass-start road events, the bike is fitted with normal dropped handlebars and ultra-light low-profile wheels.

The international cycling authority, the Union Cyclist International (UCI) insists on dual back brakes in Paralympic and international trike competition, so there is a disc brake on each rear wheel; both are operated via a standard rear-brake lever.

In addition to the double rear-brake system, the UCI also insists on a rear safety bar being fitted to all trikes in international competition, to prevent participants following each other too closely. Placing the front wheel between the two rear wheels of the trike in front is aerodynamically advantageous but potentially dangerous should the lead rider swerve or alter direction. Safety bars make this impossible.

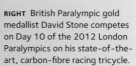

RIGHT British Paralympic gold medallist David Stone competes on Day 10 of the 2012 London Paralympics on his state-of-the-art, carbon-fibre racing tricycle.

One Rider or Two? >

'But you'll look sweet upon the seat of a bicycle made for two...' So went the last line of the song Daisy Bell, written by Harry Dacre in 1892. His song was a big hit and proved ample indication that tandems would quickly follow the solo bike's popularity. Tandems are sociable – two people ride at the same pace to the same place, sharing the effort and experience. Tandems are not simply two bikes stuck together; they have their own engineering and their own terminology.

The front rider is called the pilot while the rear one is known as the stoker. Pilots pedal, steer and shift gears; stokers just pedal. Two sets of legs are more powerful than one, and two fit riders on a tandem will travel faster than one on a bike, except perhaps uphill, where tandem riders are limited by having to stay in the saddle because it is difficult to climb out of the saddle with the required harmony.

Speed was the reason why cycle racing was and still is exciting. Once racing became established on roads and tracks, tandems were used to pace individual competitors. And if four legs are faster than two, just think how fast a five person-powered bike could go. As paced racing progressed, track promoters used three-, four- and five-seat bikes to make racing even more spectacular.

RIGHT The Raleigh Pioneer Adventurer, a modern tandem that is suitable for use over many different terrains. Its frame is constructed from aluminium and the bike has fat, semi-slick tyres for a comfortable ride. Disc brakes give good stopping power [1]. The bike provides a wide spread of gear ratios (tandem riders need lower gears than solo cyclists when climbing steep hills because it is more difficult for them to climb out of the saddle). Both sets of handlebars are flat, giving both pilot and stoker a comfortable, upright riding position [2]. The bike has 24-speed click-shift gears, with shifters mounted on the front handlebars just under the brake levers, so the pilot controls gear selection and braking [3].

Beefed up a bit >

No matter what material a tandem frame is constructed from, it is always of a heavier-grade than a solo bike frame. Tandem forks must be particularly strong because slowing and stopping a tandem puts heavy loads on the front of the bike.

In the early days, road tandem brakes were the same as whatever was being used on solo bikes, but, from the mid-1930s, rim brakes were sometimes used on tandems in combination with drum brakes. This prevented heat build-up in the rims when controlling speed on long downhill sections. Modern disc brakes are perfect for tandems.

Tandem wheels are also stronger and heavier versions of solo wheels, with heavier tyres, which are usually inflated to higher pressures than a solo rider would use. Extra spokes are used, and often tandems have wider rear-wheel axles. This is to suit a technique used in building tandem wheels. To accommodate a bike's gears, standard rear wheels are asymmetric, with shorter spokes on the drive side, which weakens the wheel slightly. This is not a problem on a bicycle but it can be on a tandem. So equal-sized spokes are used on tandems and the wider axle accommodates the gears. Modern disc wheels or thick-spoked carbon-fibre wheels make wider axles unnecessary on race tandems.

Race tandems >

Once motorbikes replaced multi-seat bikes for pacing riders, tandem sprinting became the big event. There was even an Olympic title for tandems sprinters – the last one being in 1972. The sight of two teams thundering flat out around a banked velodrome was dramatic.

International tandem racing is now found only in para-cycling and in the Paralympics, where a tandem is the only way talented visually impaired cyclists can compete. Those from the top cycling nations ride state-of-the-art machines.

Able-bodied enthusiasts meanwhile compete on tandems in road races or in time trials all over the world, but only in Britain is there still a national tandem sprint title. There is an even smaller core of racers who combine two seats with three wheels and ride what is a modern-day equivalent of a Roman chariot – the tandem tricycle.

Some tandems allow independent pedalling, but the majority do not. In the most common tandem drive-train, the pilot's legs are connected to the stoker's by a timing chain running on equal-sized chain-rings on the left side of the tandem. The stoker then has a normal chain-set on the right, which is connected to the rear wheel by a chain, just like a standard bike drive-train.

ABOVE In this tandem sprint race in the 1972 Munich Olympic Games, the Belgian team is following the West Germans in the early stages of a heat before they build up speed and go flat out for the line.

LEFT In modern Paralympic tandem racing, which takes place on the road and on the track, the tandem pilot rides with a visually impaired stoker. Here stoker David Stone of Great Britain competes in the London 2012 Paralympic Games.

Drinks on the Go >

Imagine slogging up a mountain road under a burning sun, miles from the next aid station or café and you are thirsty – very thirsty. Right from the start of cycling as a sport, racers developed ways of carrying drinks. However, in the early days of cycle racing right up until the late 1960s, rules dictated that riders could only take two bottles of drink on their bikes. Athletes trained themselves to race while thirsty.

For the first Tour de France in 1903, riders fastened leather satchels to their handlebars. Inside them they carried bottles of water, coffee, tea and even wine. Many of them carried flasks of brandy for moments when matters got really trying.

The Tour organizers set up feeding stations along the route but distances between each one varied, so riders stopped at roadside cafés, too. This was frowned upon by the organizers but, with stages in excess of 300km (186 miles), café raids persisted until the end of the 1960s.

Leather satchels began to disappear after the First World War and were replaced by cork-stopped metal bottles that were carried in special cages mounted on a bike's handlebars, with two bottles being placed side by side. Later, in the 1930s, the French rider René Vietto was the first to mount a bottle cage on the down tube of his bike.

Although others copied Vietto, it was still common to see bikes with handlebar and frame-mounted bottle cages right through the 1960s, when squeezable plastic bottles came on the scene. Far superior to their rigid metal predecessors, the cylindrical plastic bottle is still the standard form today.

Ease of use >

Cycling-specific drinks bottles often have a plastic cap with a nozzle. The nozzle has a small plastic stopper, which can be pulled out by a rider's teeth. Riders then squeeze some of the drink into their mouths. The stopper does not come right out of the nozzle, so the rider pushes it back after drinking. With practice, this whole operation can be done smoothly, in seconds, without stopping pedalling.

TOP/ABOVE Early Tour de France riders carried liquids in glass bottles stowed in leather satchels. After the First World War, they started using special flasks called *bidons*, which were carried in handlebar-mounted cages.

RIGHT The *bidon* developed into the modern drinks bottle that racers carry on their bikes today in frame-mounted cages, which are made from plastic and have stoppered nozzles. Riders squeeze the liquid through the nozzles into their mouths.

For the first Tour de France, in 1903, riders fastened leather satchels to their handlebars. Inside they carried bottles of water, coffee, tea end even wine. Many of them carried flasks of brandy for when matters got really trying.

The handlebar bottle cage has been out of fashion since the 1960s, except with triathletes, who use aerodynamic-shaped bottles attached – often horizontally – to their handlebars. Fitted with a plastic tube, the rider does not even have to remove the bottle to take a drink. Most cyclists fit two bottle cages to their bikes – one on the down-tube and one on the seat tube. Some triathletes and time triallists use behind-the-saddle cages, which remove bottles from the frontal airflow and so improve aerodynamics. Others place aerodynamically shaped bottles on their bike's down-tube or seat tube.

Drinks bladders >

Bike drinks bottles are one of those products that reached design maturity early. The only real alternative is a drinks bladder. These are either slung under the saddle or carried in small bags on a rider's back as part of a rucksack. Some bike manufacturers have experimented with placing drinks bladders inside the bike's frame itself.

BELOW The drinks container on this aerodynamic time-trial bike is triangle-shaped and held in the space between the lower part of the seat-tube and the down-tube. The rider reaches down and lifts it out of its holder to take a drink.

ABOVE As it is difficult to handle a bottle when riding over rough terrain, mountain bikers often use cycling bladders. The tube that this rider drinks from leads from the bladder inside his backpack and can be seen on his left shoulder.

Cycling Baggage >

Racers travel light, but cycling commuters and those who use their bikes for touring or adventure need to carry work-related materials, spare clothing, food and even a tent, sleeping bag and cooking utensils. These are carried on the bike itself, as cycling with anything heavier than a light rucksack on your back is not only very tiring and uncomfortable but also upsets a rider's balance.

For day rides or commutes, there is a sharp contrast between the way British and European cyclists began carrying luggage on their bikes. The Europeans have always preferred handlebar-mounted bags, while the British opted for saddlebags. Handlebar-bag proponents argue that a front-mounted load affects a bike's balance less than one mounted at the rear. Both bags vary in size depending on what carrying capacity a rider needs and bigger bags can be mounted on metal supports.

The European handlebar bag, which was perfected around the 1930s, is the Solange, and they are still available today from manufacturers such as Gilles Berthoud. In the early 1930s, Wilf Carradice came up with the classic British saddlebag in Nelson,

Lancashire, and his design persisted for many years. You can buy a version of it today from the company that Carradice founded. It is still based in Nelson and makes a range of cycling bags and luggage from tough duck cotton. The original saddlebags were attached to metal bag loops, which were integral parts of some saddles, and still appear on leather Brooks saddles today.

Handlebar-mounted wicker baskets were the first bike 'bags' that were used for shopping, but the advent of small-wheeled bikes such as the Moulton (see pages 108–9), the Raleigh Shopper and similar bikes made by European manufacturers such as Peugeot, saw the baskets replaced by rectangular box bags that were mounted on carriers over the rear wheel.

Elsewhere in the world carrying goods on bicycles is a day to day necessity. The bikes – and those who ride them – are extremely adaptable in this respect. Goods are carried using home-made slings and bags, mounted on home-made carriers, or just tied to the bike. Many bikes that are built specifically for use in Africa and in other countries, such as the Buffalo bicycle (see pages 262–3) have luggage and goods carrying-capacity built into their design.

BELOW Bicycles, complete with front and rear panniers, on forest trails of the Isar Cycle Route, Strasslach-Dingharting in Upper Bavaria, Germany.

BELOW Bicycles can be adapted in a number of different ways to carry a wide range of goods.

RIGHT A Red or Dead Starstruck bicycle with a box bag mounted on a carrier over the rear wheel.

For day rides or commutes, there is a sharp contrast between the way British and European cyclists began carrying luggage on their bikes. The Europeans have always preferred handlebar-mounted bags, while the British opted for saddlebags.

Adventure cycling >

Cyclists who need to carry more than they can fit into one bag might choose to use a handlebar bag and a saddlebag concurrently. After that they need panniers. Bike panniers are similar to the pannier bags that have been used on beasts of burden for centuries. The first patent for a bicycle pannier was taken out in 1884 by John D Wood of Camden, New Jersey, while modern bike panniers were invented in 1971 by Hartley Alley in Boulder, Colorado.

Bike panniers fit either side of the rear and/or front wheel and are attached to, and supported by, a metal frame that bolts on to the bike frame. The advantage of panniers is that they lower the centre of gravity of both rider and bike, thereby making the bike more stable.

Pannier capacity is measured in litres. An adventurous cyclist can carry all he or she needs for extensive expeditions. Mark Beaumont rode around the world in less than 80 days, carrying everything he needed, apart from fresh food, using panniers.

LEFT Explorer Mark Beaumont carried all he needed during his attempt to circumnavigate the world by bicycle. He is seen here finishing his epic ride in Paris in February 2008, having completed 18,000 miles in just 195 days.

LEFT A fully loaded touring bike with an extra bottle mount below the down tube, front and rear panniers, an extra bag mounted on top of the rear pannier rack and a handlebar bag.

The Evolution of Children's Bikes >

The bicycle really is the gift that keeps on giving. Our adult health and fitness are partly determined by what we did when we were young. Evidence suggests that an active, fit child grows up to be an active, fit adult. In addition to the health benefits, children have traditionally used bikes to explore where they live, expand their horizons and begin to understand the concept of travel.

The first bikes that were made specifically for children are still around today and they are an everyday demonstration of how high-wheeled bikes worked. They are tricycles that are designed to be ridden by one to three year olds and are equipped with pedals that are attached directly to their front wheels. They have two smaller wheels at the back for balance and their design has changed very little in the last 150 years. You can even buy these tricycles made out of wood, just as the first kids' bikes were.

Tricycles continued to be the only mass-produced bikes available to youngsters in Europe until after the Second World War, although they were metal-framed, chain-driven trikes, bigger than those intended for the very young. Two-wheeled children's bikes developed much earlier in America where, following the initial bike boom of the 1890s when Americans took to driving cars, the bicycle was seen either as a toy or as transport for the young.

It was a different story in places around the world where cars were not affordable or were not a practical form of transport. Bicycles were used by people of all ages in countries such as India and China, where they still play an enormous role in transporting people and goods.

Smaller versions of adult bikes went into mass production in Europe after the Second World War. As the 20th century progressed, the size of children's bikes was determined by the bike's wheel size: 12-in, 16-in and 18-in wheels for younger children, then 20-in and 24-in wheels for children between 9 and 12 years of age. Once the 24-in stage is outgrown, a child is large enough to graduate to an adult bike.

ABOVE A child's cycle with the pedals attached to the front wheel, which gives the same direct drive as the ordinary bicycle (see pages 16–17).

RIGHT By the 1950s and 1960s, many Americans saw bicycles as suitable toys for children. They were designed to copy the form of motor cycles or even motor cars.

Two-wheeled children's bikes developed much earlier in America, where, after the initial bike boom of the 1890s and Americans took to driving cars, the bicycle was seen as either a toy or as transport for the young.

BMX Bikes >

The advent of bicycle motocross (BMX) in the 1970s encouraged kids to get on bikes, and it still does. BMX has grown as a sport since it began and now features in the Olympic Games (see pages 154–5). Early BMX bikes had spoked wheels, but plastic Mag wheels were introduced during the late 1980s and early 1990s. Later, most reverted back to spoked wheels. Today's BMX bikes have chromoly frames and some specialized parts, such as pegs for standing on to do street tricks. Mountain bikes continued the momentum started by BMX (see pages 136–7).

When they first became widely available, BMX and mountain bikes were revolutionary – a quality that always appeals to the young. Both bikes still have a youthful image. In the early 1990s, the production of mountain bikes saved many bike manufacturers from bankruptcy and exposure to mountain bikes has helped thousands of young people to discover cycling.

ABOVE An early junior-category BMX race on a prepared track. BMX racing became an Olympic sport in 2008 and its champions are among the fastest cyclists on the planet.

ABOVE/RIGHT A BMX bike set up for performing tricks, with a device on the bike's headset that allows the front wheel to turn through 360 degrees independently of the brake cables [1], and stunt pegs attached to the rear wheel nuts so the rider can stand on them [2].

Bikes at Work >

From the delivery bike with its big basket on the front to ice-cream vendor bikes and pedal taxis, bikes have been used for work purposes ever since they were invented. They still have a role to play in the workplace – one that is certain to grow in the coming years.

Butcher's bikes, baker's bikes, candlestick maker's bikes. Well, no, perhaps not candlestick makers. I have no direct evidence that candlesticks were delivered by bike but some probably were. Right up until the 1960s, and later in some places, delivery bikes were how most household items were moved between shop and home. It is how the post is still delivered in many countries.

Delivery bikes have a small front wheel and normal-sized rear. The front wheel supports a big metal carrier secured to the bike's frame. Anything can go in the carrier, from custom-made wicker baskets to boxes or mail bags. In cities where bikes are part of the transport infrastructure, such as in the Netherlands, delivery bikes are still used a lot.

There are many other bikes doing a job of work. Tricycles with large flat-bed carriers low down behind their saddles and tricycle taxis of all shapes and sizes carry people in towns and cities around world. Tricycles with two front wheels bear ice-cream vendors and their wares. And what about the beer bike? This Dutch idea comprises a mobile bar that you hire as a party, sit at and pedal along while one of you drives and a person from the hire company serves you beer. There are 44 of them available to hire in the Netherlands. You can ride where you want – around the streets or along country lanes.

Special place >
Bikes have a special place in less developed countries, where ordinary bikes are pressed into service to carry all manner of goods. Bikes have also been adapted to drill water boreholes and supply electricity, as well as to serve many other useful purposes. Charities behind much of this work depend on donations of unwanted bikes that can be adapted and pressed into service again.

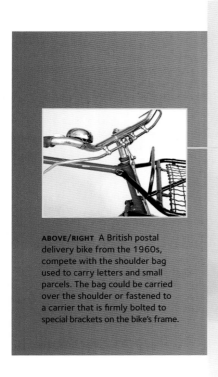

ABOVE/RIGHT A British postal delivery bike from the 1960s, compete with the shoulder bag used to carry letters and small parcels. The bag could be carried over the shoulder or fastened to a carrier that is firmly bolted to special brackets on the bike's frame.

Bikes have a special place in less developed countries, where ordinary bikes are pressed into service to carry all manner of goods. Bikes have also been adapted to drill water boreholes and supply electricity, as well as to serve many other useful purposes.

Bike messengers >

One group of working cyclists makes a big impression on the world's streets today: bicycle messengers. They have their own culture; and even helped create a new bike – or at least a new name for one that already existed, the Fixie (see pages 230–31). But bike messengers are not new.

Bike couriers were employed by the Paris Stock Exchange in the 1870s, and during the 1890s American cycle boom there were bike couriers working for the Western Union and the Telegraph Office. They delivered telegrams for the British and Australian post offices and did a similar job throughout Europe.

The current wave of bike messengers started in San Francisco in the 1980s but spread rapidly when businesses realized that such messengers could deliver quickly in traffic clogged cities any documents that needed physically signing.

Such messengers often use old bikes, and they particularly like fixed-gear steel-framed track bikes. Messengers' clothing is a mix of practical and personal but there are two must-have items: a large courier bag slung across their backs and a two-way radio clipped to the bag's thick shoulder strap.

TOP LEFT Bicycle messengers with the Western Union working the streets of Seattle in 1937. Theirs is an old profession that continues to thrive in a digital world.

ABOVE/RIGHT Bicycle couriers are a familiar sight on streets throughout the world. As well as delivering messages, bikes can transport all sorts of goods, as the picture on the right so eloquently shows.

Bikes at War >

In 1924, Ottavio Bottecchia became the first Italian to win the Tour de France, even though he had not ridden a bike until the First World War. Bottecchia discovered his talent in the Austrian army, when he was a bicycle-mounted machine gunner. Dodging bullets must have been good training, because when Bottecchia got out of the army he started racing. Shortly after he became a professional and started winning big races.

Bikes enabled troops to move more quickly than they would on foot; they were also easier to care for than horses, and less conspicuous. In 1885, the Austrian army established a bicycle cavalry with its headquarters in Vienna, where riders were taught to fight with swords and guns while cycling.

The British army used cyclists for reconnaissance and communications during the Second Boer War (1899–1902). Then, in 1908, they established nine battalions of cyclists in the Territorial Army, which grew to 14 by the eve of the First World War. Initially, the cyclists were used for coastal defence duties but in 1915 the Army Cyclist's Corps was founded and two battalions – the Kent Cyclists and the Royal Scots – saw service on two fronts in a reconnaissance role.

ABOVE An example of the Swiss Army bike from 1991. Bikes make adaptable transport, and if the terrain becomes too rough to ride, you just get off and push. That fact made bikes perfect for moving troops around.

ABOVE A convoy of bike-mounted troops on the roads of northern France during the First World War.

ABOVE Members of the fully equipped Bersaglieri (Italian rifle battalion) move cross-country with their bicycles in 1915. Bike-mounted troops travelled faster than infantry and performed a similar function.

The Germans used bikes in a similar way but the Italian Bersaglieri took matters further. When fighting in the Alps the troops carried early versions of fold-up bikes so they could move faster when conditions allowed.

Nearly every army involved in the Second World War had cycling groups for both defensive and offensive purposes. Allied paratroopers were supplied with folding bikes to help them move quickly once they landed; in doing so they helped the Allied invasion to end the conflict. Armies also adapted bikes so that they could provide power to operate radios.

During the Vietnam War (1954–1975), the Viet Cong used bicycles to ferry supplies down the Ho Chi Minh Trail. Individual riders carried loads of up to 180kg (400lb). Cyclists were more difficult to see from the air and bikes are easier to get off the road and take cover with. Using bikes, the Viet Cong avoided the air attacks that motor transport would have attracted.

The Swiss army bike >

This is the ultimate battle bike. The first model was used by the Swiss army in 1909 and improved until the 1990s. It would be nice to think that the bike had all sorts of weapons and other kit stashed inside its tubes but it has not. The bike's main characteristics are its considerable durability and its very large luggage capacity – a bag fills the whole of the main triangle of the bike's frame.

In 2012, a new Swiss army bike – the MO-12 – was introduced. This had a black aluminium frame, eight-speed Shimano Alfine hub gears with a hub dynamo, Magura disc brakes and a custom tool bag. At the time of going to press, civilians can buy them for 2,795 Swiss francs.

ABOVE/LEFT This design of Swiss Army bicycle, the MO-5, which dates from 1905, was replaced by the M-93 in 1993, and that was used until 1995. Details include the rear hub brake [1], a leather satchel stowed between the seat tube and seat stays [2], handlebars showing the rod front brake and bell [3] and a dynamo powered by the spin of the front wheel [4].

Chapter 5
Fighting the Wind >

Racers at speed on a velodrome. The rider in second place is in the slipstream of the front rider and is therefore expending significantly less energy than the front rider. Track racers like these were the first cyclists to understand that aerodynamic drag was the biggest factor to be overcome in the search for increased speed.

Sprinters and Stayers >

The biggest factor, other than the rider's fitness, that limits how fast a cyclist can go is air resistance. For cyclists and their bikes, air resistance is proportional to the square of their velocity, so doubling velocity produces approximately four times the resistance. The faster you go, the worse it gets as resistance grows exponentially. To put it into context, approximately 80 per cent of a racing cyclist's energy output is being used simply to push air out of the way.

Track racing was originally the preserve of high-wheelers (see pages 16–17), but as the safety bicycle (see pages 24–5) became dominant and speeds increased, banked velodromes had to be built to keep the riders safe. The advent of such higher speeds and bankings required new tactics, so riders started to think about ways of using the air resistance they encountered to help them win.

In the early days of track racing, crowds flocked to meetings mainly to watch the match sprints. In these races, two riders competed against each other in heats, and the winner of each heat progressed through further rounds to a final. The increased air resistance from the faster safety bikes meant these racers were no longer 'flat out from gun to tape' affairs, as seen at athletics meets. Instead, each rider used his or her opponent's slipstream to saves energy – as much as 40 per cent – before making a late charge for the line.

For this reason, cat-and-mouse-style tactics evolved, with each sprinter trying to force his or her opponent to take the lead until the sprint proper started, and each would try to slingshot by in the final few metres (yards).

Through the early 1900s, the exploits of sprinters Thorwald Ellegard of Denmark and Arthur Zimmerman of America caused people to pack into velodromes in Europe and the USA. As they began to understand the relationship between their upright

BELOW LEFT Top Dutch professional sprinters (both in dressing gowns) – Jan Derksen (left) and Arie Van Vliet (right) – talk to fans between rounds of the Réunion au Vélodrome d'Hiver meeting in 1942.

BELOW RIGHT In match sprints on wide outdoor tracks, cat-and-mouse tactics were crucial. The rider who led out the sprint was at a disadvantage because the other could slingshot past out of his slipstream. Here Arie Van Vliet (right) narrowly defeats his French rival, Louis Gérardin at the Parc des Princes velodrome in 1948.

OPPOSITE LEFT The first paced races were paced by riders on tandems or triplets. The competing rider would sit in his pacer's slipstream, where he could ride much faster than normally. The man being paced here is Charles Stewart Rolls, founder with Sir Frederick Henry Royce of the Rolls-Royce car company.

OPPOSITE RIGHT Modern paced races include the keirin, where a small motor bike paces the riders for a number of laps, then moves out of the way and the riders sprint for the line. Keirin racing originated in Japan, where it is a hugely popular sport.

riding position and air resistance – and therefore speed – sprinters such as Ellegard and Zimmermann began to adopt crouched positions and dropped handlebars, and these features became the norm among sprinters as they raced flat out around the track. This rudimentary understanding of aerodynamic cause-and-effect changed both tactics and positions. With the move to banked tracks to manage the increased speeds came

further tactical changes. Sprinters started launching their attacks from the top of the new banked corners, swooping down to boost their acceleration.

With tracks now capable of accommodating faster racing, a new kind of event was born. Paced distance racing and competitors were split into two categories – sprinters and stayers – like greyhounds are today.

Paced racing >

Stayer races were long, and the spectacle might have palled if track promoters had not come up with a way of making these endurance events faster and more entertaining. Initially, they used tandems to pace individual competitors, then multi-seated bikes. Companies such as Dunlop employed pace teams, which would be swapped several times during a 160-km (100-mile) race so they stayed fresh to pace their star rider on his solo bike as fast as possible.

Eventually, the pace teams were replaced by motorbikes. The first pacing motorbikes had two riders – one to steer, and one who sat upright at the back to provide shelter. Soon, the motorbikes were adapted with long handlebars so the back rider – standing to provide maximum shelter for his charge – could steer, and so the front man was no longer needed.

These races were very fast, with average speeds of 96km/h (60mph) and more. The sheltering effect of the motor pacer meant riders could keep going at this pace for 80–160km (50–100 miles). Tracks for this sort of racing had to have steeper bankings than normal.

With ten or more riders and the same number of noisy motorbikes thundering around the tracks at incredible speeds, the atmosphere must have been incredible. It was also very dangerous. Crashes at this speed were often life-threatening, so rollers on a protruding metal frame were fixed behind each motorbike to prevent the racers following too closely. The resulting drop in speed improved the safety record, but it remained much more hazardous than unpaced racing.

Motor pacing is still called stayer racing in Europe, and a world championships for this very special race persisted until 1992.

The Need for Speed >

Human beings are not one of nature's aerodynamic success stories. Even when perched atop a sleek machine, their arms and head stick out while the legs thrash up and down disturbing the airflow. In other areas of the natural world, certain creatures – fish and birds, for example – are far more aerodynamic, so some pioneering cyclists decided to see if they could apply some of the natural world's features to cyclists, by using fairings.

A number of inventors came up with aerodynamic covers that smoothed out the airflow over cyclist and bike. One of the most effective was developed in Manchester; it was called the Sputnik after the Russian satellite.

The Sputnik bike was based on the home-made fairings that some cyclists started attaching to their bikes in the 1920s and 1930s. The idea was that a bike and rider streamlined by a fairing would have an advantage in races where aerodynamics were key, such as the individual time trial and the individual pursuit. It would also work in record attempts, which are done alone and unpaced.

The Sputnik was a fairing that was developed by members of the Manchester Wheelers Cycling Club and it could be attached to any bike. It was first presented at a meeting that was promoted by the Manchester Wheelers at the city's Fallowfield track in 1957. It was ridden by a Frenchman, François Lahaye, in a special 4,000-m (2½-mile) pursuit match in which he beat a first-class team pursuit squad made up of Norman Sheil, John Geddes, Mike Gambrill and John Plumbley. In such a pursuit event, two riders or two teams of riders start on opposite sides of the track and try to catch each other inside the 4,000-m (2½-mile) Olympic distance. This seldom happens, so the winner is whoever covers the distance in the shortest time.

As a rule, a top-level pursuit team is faster than a top-level individual in a pursuit event because in teams members swap the lead and share the workload, thereby conserving energy. Although Lahaye was a good racer, the team that he was up against was in another league altogether. It contained the 1955 world individual pursuit champion (Norman Sheil), plus two of Great Britain's 1956 Olympic team pursuit bronze medallists (John Geddes and Mike Gambrill). Yet, Lahaye beat them easily.

The Sputnik bike is now on permanent display inside the Manchester Velodrome.

RIGHT A 1960s racer tries out some of the latest aerodynamic ideas around at the time, including a see-through fairing, a plastic crash-hat cover and disc wheels on a Moulton bike. Fairings like this were never officially allowed in racing, and the disc wheels were years ahead of their time.

A number of inventors came up with aerodynamic covers that smoothed out the airflow over cyclist and bike. One of the most effective was developed in Manchester; it was called the Sputnik after the Russian satellite.

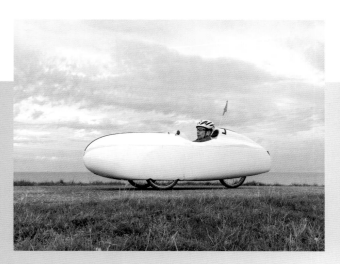

ABOVE The aerodynamic improvements over a standard bike are obvious on this HPV (Human-Powered Vehicle). The rider sits recumbent, so the frontal area of the bike is low, and the vehicle body smooths the air flowing over it.

ABOVE A selection of HPVs ready for a race in Bernkastel-Kues, Germany, in July 2013.

Fairings for leisure cycling >

In mainstream competition, fairings are not allowed, but leisure cyclists can certainly taken advantage of them, both to travel faster and to protect themselves from bad weather.

They are often made from textiles stretched over a frame, and several combinations of front, rear and complete fairings are available. Front and rear fairings work best on long bikes, such as recumbent or tandems, but a cyclist on a normal bicycle can also benefit by adding a complete fairing, known as a body sock.

Fairings might be of benefit to leisure cyclists, but any addition to a race bike that improves airflow and is not part of the structure of the bike is not allowed in competition.

Rules on design >

In the 1990s Graeme Obree and myself implemented several aerodynamic strategies involving both position and equipment – so much so that a raft of very complicated rules was hurriedly brought in to keep cycle sport true to its roots as a test of physical rather than mechanical prowess. The cycling authorities now stipulate the frame's shape (it must be a double-diamond-shaped collection of tubes), the ratio of a frame tube's depth to width and even how much the bike can weigh. Manufacturers are now constantly innovating to the limit of these new boundaries as they are all too aware of the possible gains improved aerodynamics can bring to performance – and by extension sales.

Another Way >

How well a cyclist slips through the air depends on three main factors. The first is the bike and the rider's combined frontal area – or, more crudely, the size of hole they both punch through the air. The second factor is how smoothly the air flows over the rider and bike. And the third one is how tidy the air left behind is. The frontal area can be affected a lot by having the rider in a prone position. Recumbent bikes – usually placing the rider in a near-prone, feet-first position – drastically reduce the amount of air they must push out of the way, enabling much higher speeds to be reached with the same power output.

The Velocar, a four-wheeled vehicle for adults that was pedalled – not powered – by a motor, was invented in France by Charles Mochet in 1930. Velocars were designed to be aerodynamically efficient and fast. Mochet also tried to enter the bicycle market in 1931 by cutting a Velocar in half to make a two-wheeled version. It had two 50-cm (20-in) diameter wheels, a wheelbase

of 146cm (57in) and cranks that were 12cm (5in) above the seat, so riders pedalled with their legs out in front of them. There had been recumbent bikes before, but Mochet publicized his by going for records.

In 1932, Mochet approached the less famous of the two Faure brothers, Francis, and asked him to ride in his 'velo-couche' on the track. Faure then applied to the Union Cyclist International (UCI) to attack the world hour record. The UCI responded that, since there were no aerodynamic fairings in the design, he was free to do so. Consequently, in 1933, Francis Faure covered 45.055km (27.9 miles) in an hour on a track in Paris, smashing the 44.247-km (27.4-mile) mark of Switzerland's Oscar Egg that had stood for nearly 20 years.

What followed was a story that would be replayed some 40 years later by Graeme Obree and myself. Maurice Richard also beat Egg's record with a distance of 44.777km (27.9 miles) in 1933 – but on a conventional upright bike. The UCI had a problem, which record should they ratify?

RIGHT Prospective purchasers inspect Velocars in Paris. They never quite became the revolution in personal transport that their inventor Charles Mochet hoped they would, but they do play a part in the story of the bike.

The Velocar, a four-wheeled vehicle for adults that was pedalled – not powered – by a motor, was invented in France by Charles Mochet in 1930. Velocars were designed to be aerodynamically efficient and fast.

Some of the UCI's decision-makers saw the recumbent as the future of cycling; others didn't even think it was a bicycle and argued that Faure had won little when he was not riding a recumbent, but that Richard had. They won, and Richard's record was ratified. They then set about defining what a bicycle was for recognized competition. It was an important moment precipitated by free thinkers, using their brains as well as their legs to affect the outcome. From this point on, competition cycling was defined as an upright activity: riders would pedal with their legs below them and their handlebars in front.

Revolutionary thinker >

This definition of competition cycling proved sufficient until the 1990s, when another revolutionary thinker, unconstrained by notions of what a bicycle 'should' look like, attacked the hour record. His name was Graeme Obree, and the response of the UCI to his radical interpretation of the rules was similar to that encountered by Faure (see pages 204–5).

Modern recumbent >

Recumbent bikes now come in all shapes and sizes. There are recumbents for touring, recumbents for racing against other recumbents, recumbent tandems and recumbent trikes. Hand-cycles, which feature in the Paralympics, also fall into this category. A recumbent's low centre of gravity makes it safer than an upright bike, and with full lumber support, there is no doubting its comfort. However, recumbents have a much wider turning circle than upright bikes, are more difficult to ride uphill, and are harder to spot in traffic.

ABOVE Modern recumbent bikes are perfect for relaxed touring rides. They are not allowed to compete in races with standard bikes.

LEFT Francis Faure on his horizontal bicycle in France in 1934. Recumbent bikes were so successful at setting records that they threatened to change the way bicycles looked forever.

Human-Powered Vehicles >

The Union Cyclist International (UCI) did not simply throw out Francis Faure's record and the Velocar idea (see pages 80–1), but also established a new category – that of records set by human-powered vehicles (HPVs). In doing so it set up another sport – HPV racing. This is just like cycle racing, with group races and established records. Because of an HPV's inherently small frontal area and use of fairings, the speeds reached far exceed those seen in conventional bicycle racing.

HPV racing is particularly popular in Australia, with its HPV Super Series of 6- and 24-hour races open to teams from schools and communities, each designing their own vehicle. The competitors race on short street circuits of around 1.6km (1 mile). Despite the short nature of the courses and the abundance of acute-angled turns, top competitors are able to ride distances in excess of 240km (150 miles) during 6-hour races.

World HPV records are overseen by the International Human-Powered Vehicle Association (IHPVA). Recognized records include a land speed record, which is a measured average speed over 200m (218 yards), with a flying start. The current solo record is 133.78km/h (83.13 mph) set by Sebastiaan Bowier at Battle Mountain in 2013, which is the fastest any unpaced cyclist has ever gone on the flat.

Records are only ratified if set along a dead flat straight, a bit like motor-powered land speed records. There are records for the different sexes and age groups, in legs- or arms-only categories, with single or multiple riders providing the power.

RIGHT Competitors line up before the start of the World Human Powered Speed Challenge at Battle Mountain, Nevada, with insets of some of the action.

Sebastiaan Bowier >

Bowier is Dutch. His world record was set on a recumbent bike called VeloX3. It was designed and built by the University of Delft and was covered with a carbon-fibre shell. It was coated with the same super-smooth finish given to Formula 1 cars and so experienced just one-tenth of the air resistance of an upright cyclist – that is, about the same air resistance as a car's wing mirror. Bowier's position was so extreme that he navigated using a TV camera, and because of the high ratio gear he had to pedal on the VeloX3, to exceed130km/h (80mph) he required 8km (5 miles) just to reach maximum speed.

ABOVE Some of the brightest minds and the most innovative technology are involved in HPV racing, resulting in efficient machines and phenomenal speeds.

ABOVE The recumbent bike VeloX3 was the creation of students at Delft University of Technology and VU University Amsterdam. Here it is being ridden on the A31 road in the Netherlands, where it achieved a top speed of 78.8km/h (48.9mph).

ABOVE The bicycle makes human-powered flight not only possible, but beautiful. This man-powered aircraft soars above the English Channel on a crossing to France.

The flying cyclist >

HPVs are not just restricted to riding on land. If you consider what men and women have ever achieved in transport (apart from space flight) somebody will probably have done it on an HPV.

Deadalus 88 was a human-powered aeroplane developed at the Massachusetts Institute of Technology (MIT), and in 1988 Kanellos Kanellopoulos pedalled it to fly 119km (74 miles) across the Aegean Sea. There are even pedal-powered balloons. Then, AeroVelo (a group of Canadian students) won the Igor I. Sikorsky Human Powered Helicopter Competition when they flew their human-powered helicopter named Atlas some 3.3m (11.8ft) off the floor for 64 seconds in 2013.

Cycling on water has also featured at the IHPVA speed championships, where records such as Bowier's are set and ratified. In 1987, human-powered watercraft took a giant leap when Allan Abbott propelled a hydrofoil called Flying Fish to a speed of 12.94 knots (14.89mph). MIT's Decavitator, ridden by Mark Drela, achieved a speed of 18.50 knots (21.3mph) in 1991, but he missed out on the DuPont Speed Prize for the first to break the 20-knot (23-mph) barrier.

Land Speed Records >

Human-powered vehicles (HPVs) have broken the 130km/h (80mph) barrier – 133.7km/h (83.1mph) to be precise – so HPVs are unquestionably faster than normal bikes. Yet the absolute speed record for someone peddling a bicycle is more than double this mark...

On 3 October 1995, a Dutchman, Fred Rompelberg, was recorded doing 268.831km/h (167.044mph) over a measured 1.6km (1 mile) on the Bonneville Salt Flats in Utah. Astride a very special bike and clad in full motorcycle leathers and helmet, Rompelberg was paced up to speed by a specially modified dragster motorcar.

His record-breaking machine had suspension, an extra-long wheelbase and small wheels; these were all for stability at very high speed, where even a small deviation from course could result in disaster. To be able to pedal, the rider needed a custom reduction gear system utilizing a 70 x 13 chain-ring to drive a 60 x 15 gear. This mammoth ratio carried Rompelberg 34.7m (114ft) for every pedal revolution. As the bike was impossible to pedal from a standing start, Rompelberg was towed up to speed by his pace car before being released.

In 1899, the American Charles Minthorn Murphy had become the first cyclist to ride 1.6km (1 mile) in less than a minute. The feat not unsurprisingly saw him rechristened by the public as Mile-a-Minute Murphy. His record was set pedalling behind a freight wagon of the Long Island Railway on a specially prepared length of track, on which a smooth wooden surface had been laid between the rails.

Although both of these records were set on specially prepared surfaces, normal road have also been used to set cycling speed records. In 1962, José Meiffret of France was paced by a Mercedes 300SL car to a speed of 204km/h (127mph) along a German autobahn.

LEFT Dutch cyclist Fred Rompelberg set the world land speed bike record in 1995 on this bike. He wore a motorcyclist's full leathers and a crash helmet while doing so, and his bike was equipped with a reduction gear system. He kept in touch with his pacing vehicle by radio.

ABOVE Rompelberg at speed on the Bonneville Salt Flats, Utah, behind a racing car. A specially made shelter at the back of the car channels air around the cyclist.

The gravity boys >

If you think Rompelberg's speed of 268.831km/h (167.044mph) behind a race car sounds dangerous, what about 222km/h (138mph) straight down the side of a mountain... on snow? That is what Frenchmen Éric Barone and Christian Taillefer did, both achieving the same ludicrous top speed of 222km/h (138mph) in 2000. The fact that they topped out at the same mark probably means this is, for all intents and purposes, terminal velocity for a bike and its rider while not being paced (and still remaining in contact with the Earth's surface).

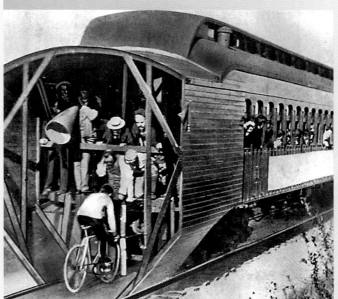

LEFT Mile-a-Minute Murphy is urged on by supporters as he attempts to be the first cyclist to ride at 96.5km/h (60mph) on a specially prepared rail track on Long Island, New York, in 1899.

ABOVE The Frenchman Éric Barone pushes the boundaries of un-paced bicycle speed by reaching 222km/h (138mph) on this icy downhill in Vars, France, in April 2005.

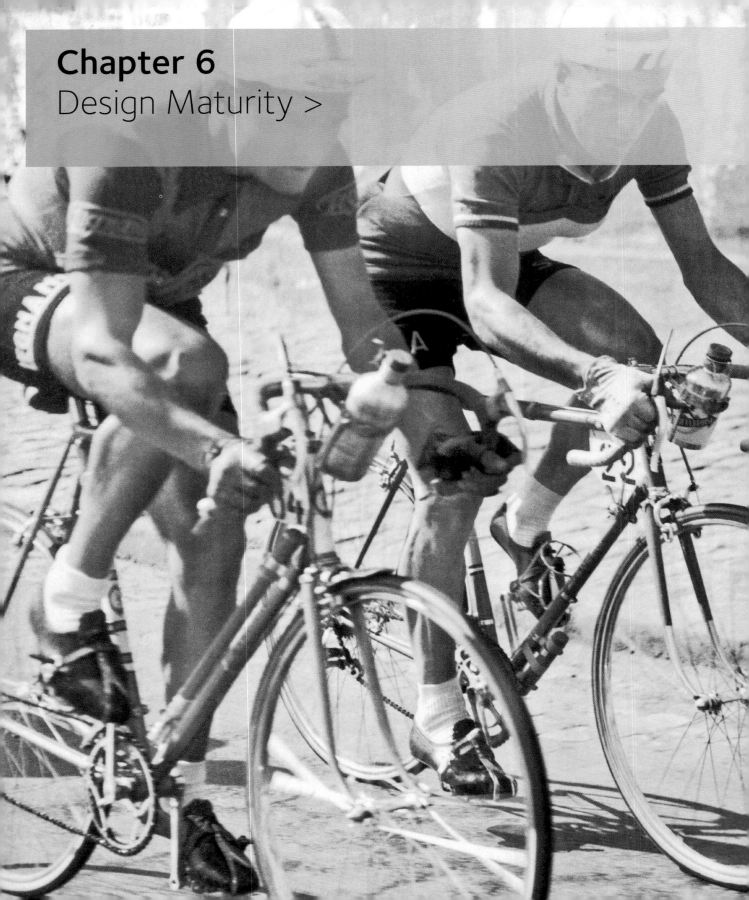

The Paris–Roubaix race in 1952. Louison Bobet of France, the first man to win the Tour de France three times in successive years, races shoulder to shoulder with the Italian, Giueseppe Minardi. Rik Van Steenberegn of Belgium won the race.

Painted Aluminium >

The double-diamond shape of today's bike frame is visible in the design of the first safety bicycle (see pages 24–5) and has remained largely unchanged since. Instead, advancements have come via small-step improvements in construction methods and materials.

In 1903, the first Tour de France bikes were made of steel and had wheels with wooden rims. The seat and head tubes had a much more backward lean than the bikes of today. The 'relaxed' frame angles helped to absorb bumps from the rough road surfaces over which racers had to compete. A modern bike with its faster handling and steeper angles would not have coped well on that early terrain and would have made the habitually long distances of the first editions of the Tour all but intolerable.

As road surfaces improved, so bicycle frame angles changed to take advantage of the faster surfaces. To increase responsiveness and improve stiffness, angles grew steeper and wheelbases grew shorter. Although these changes were hardly noticeable from year to year, if one of the bikes from the first Tour de France in 1903 is compared with a contemporary machine – or even with one from the 1930s, by which time the majority of roads were coated with materials we would recognize today – the evolution is plain to see.

Mavic magic >

Wooden wheel rims might sound archaic but they were lighter and more comfortable than the alternatives of the day. In the 1930s, French cycle component company Mavic used an aluminium alloy called Duralumin, developed by German metallurgist Alfred Wilm in 1903, which was used in the construction of the first airships. The alloy was far lighter and more responsive than wood; however, under the rules of the day, Mavic's Duralumin wheels could not be used in the Tour de France. Despite being banned, the French company persuaded the 1931 Tour de France winner Antonin Magne to use the wheels in the 1934 event. They avoided the attention of the judges by painting a wood grain over the aluminium surface. Magne won the race and the deception was revealed. Mavic and other manufacturers then lobbied the Tour organizers to allow aluminium components into the race. They agreed and their consent opened the floodgates for development. Soon handlebars, stems, chain-sets and brake callipers all had lightweight aluminium incorporated into their design.

Tourist routiers >

Literally translated, a tourist routier is a 'tourist of the road' and, until just before the Second World War, they made up

LEFT A 1903 racing bike, complete with bags for carrying food, drink and essential spares. Although this bike looks old-fashioned, it was perfectly suited to the rough roads and long-distance races of its era.

RIGHT Competitors assemble for the start of the 1927 Tour de France. Some of the stages in early Tours were so long that riders started in the dark, rode through the day and back into darkness before reaching the finish line.

OPPOSITE Antonin Magne rides to Tour de France victory in 1934 on aluminium wheels, painted to look like they were made from wood.

The largest contingent of tourist routiers in the Tour de France were bike shop owners. Post event, they would often display their machines in their respective shop windows, demonstrating the quality of their wares – able to survive the rigours of the grand race.

the majority of the early Tour de France field. The rules of the day stated that, as amateurs, they had to compete the distance wholly self-supported. They had no teams and no outside financial backing. During and after each stage, they had to supply their own food and even arrange their own overnight accommodation. Although it was undoubtedly a tough way to tackle the monumental challenge of the Tour de France – often only 10 per cent of their number made it all the way to the finish line in Paris – there was no shortage of willing volunteers, including some interesting characters. Notable among these were a Parisian policeman, Gelot, who was given the nickname Le Flic Volant (the flying copper) by the attending press. Also participating in several editions of the event was a geography teacher from Perpignan.

The largest contingent of tourist routiers in the Tour de France were bike shop owners. Post event, they would often display their machines in their respective shop windows, demonstrating the quality of their wares – able to survive the rigours of the grand race. Completing the Tour undoubtedly enhanced their own and their businesses' reputation and was a strong draw for potential customers.

Search for new materials >

Bike manufacturers were always on the lookout for materials that might perform better than the ones they were currently using. In 1895, the bicycle craze inspired Swiss industrialist Ralph Steifel to invent the seamless steel tube. This was stronger and lighter than the previously used tubes, which were formed from sheet metal, rolled and welded together. Tubes continued to be refined, with walls becoming paper-thin in a quest to reduce weight. These were left slightly thicker towards the end of each tube where they joined, to preserve strength, a process called butting. The search for new materials resulted in experiments with aluminium, carbon fibre and exotic metals such as titanium (see pages 118–19, 182 and 184). The quest for ever lighter and stronger ways to build bikes goes on to this day.

BELOW The winner of the first-ever Tour de France, Maurice Garin, poses with the winning bike at the end of the 1903 race.

The Name Game >

One of the prime reasons bike manufacturers get involved in sponsorship is the knowledge that people will want to ride if not the same model of bike, then one from the same manufacturer, as those of their sporting idols. For this reason, in its early years the Tour de France was contended by trade-sponsored professional teams and by individuals.

In 1930, however, Henri Desgrange, the then director of the Tour, decided that the race should only be disputed by national squads and individuals, not by trade-sponsored teams, as it had been up until that point and is today. Anyone wishing to compete was obliged to use an unbranded yellow bike that would be supplied to them. Desgrange introduced the rule as he felt that trade-sponsored teams would collude in the race, whereas national teams would not. The use of identical unbranded bikes removed from the equation any element of one manufacturer's bike being better than another in determining who would win. Degrange felt that these changes would make the Tour a purer race based purely on athleticism and ability.

The sponsors, who paid for teams taking part in all the other bike races on the international pro calendar, resented the Tour's high-handedness and, by the mid-1930s, a compromise was reached. The participation of national teams remained for the Tour but the riders were allowed to ride branded bikes and to carry the names of their trade team sponsors on their national colours.

Matters were far worse for bike manufacturers in Britain, and for far longer. Complaints to the authorities from members of the public following some of the earliest cycle races made British cycling body the National Cyclists' Union (NCU) nervous that the police would crack down on cycle racing and impose rules on where and when races could be held. To pre-empt this, and to avoid an all-out ban on road cycling, the NCU asked member clubs only to promote races on cycle tracks or other enclosed road facilities, such as airfields. Another cycling body, the Road Time Trial Council (RTTC) avoided drawing attention to its activities by stipulating that individual, 'private and confidential' time trials were the only racing allowed on open roads. What is more, competitors in time trials had to be clothed neck to foot in black in order to make cycling competition as inconspicuous as possible. These rules persisted from the turn of the 20th century until the 1950s.

Yet another RTTC rule could have stifled cycling in Britain if it had not been for the ingenuity of bike manufacturers. Just as with the 1930 Tour de France competitors, cyclists competing on the open road in Britain from the early 1930s were prohibited from carrying advertising on their bikes, including the names of bike manufacturers. Cycling in Britain was strictly amateur at the time and cycling authorities wanted it to stay that way; unsullied, as they saw it, by money. Yet some bike companies provided the best racers with free bikes, and some even paid competitors to ride. Suddenly, there was no point in doing so if manufacturers could not reap the benefit of people knowing which bike a particular rider raced on. So, to get round the rule, some manufacturers came up with ways to make their bikes distinctive without labelling them. Here are three of the most famous examples.

The Carlton Flyer >

In the 1930s, Carlton Cycles made its top model, the Flyer, identifiable by building it with very steep frame angles and a very short fork rake. Despite the lack of branding, its shape alone made it sufficiently recognisable for people to know the maker.

The company, which was located in the Nottinghamshire town of Worksop, supplied their distinctive bikes to great Yorkshire time triallist Harry *Shake* Earnshaw, who became synonymous with the brand.

The Hetchins >

Another manufacturer, Hetchins, whose bikes were recognizable by their elaborately curved tubes, followed the same approach, supplying products to the fastest men in the Bronte Wheelers, another Yorkshire cycling club, including the one-time 160-km (100-mile) record holder Reuben Firth.

The Baines Flying Gate >

There are a few different versions of the odd-looking but unmistakable Flying Gate frames, which were first produced in 1935. All have a vertical seat tube. Although it was called a seat tube, the bike's saddle was not in fact connected to it.

Established in 1898, Bradford-based Baines – the makers of the Flying Gate – supported two local time triallists: Jack Fancourt and Jack Holmes. Fancourt won the cycling version of the Isle of Man TT in 1937 on a Baines Flying Gate.

Ladies Only >

Until quite recently, the design of women's bikes was driven by men rather than the wishes and needs of those who rode them. Even though the desire to ride the first bikes changed women's fashion (see page 31), it was men who decided that bikes designed specifically for women should have no top tube (or a much lowered one), to accommodate skirts and dresses. That also meant that women's bikes looked very different to men's. If a woman wanted to race, she had to buy a man's bike or have one custom-made for her, which was expensive.

This didn't change until the 1990s when Women's Specific Designs (WSD) became commonly available. WSD bikes look and handle in a similar way to men's bikes. The differences are subtle and take into account the morphological differences between men and women: frame sizes are smaller, bars are narrower and cranks are shorter.

Classic design >

Classic bike designs for women were loop-framed (similar to those commonly used today by both sexes in the Netherlands) and date back to almost the beginning of cycling. The mixte frame, which was introduced mid 20th century and still had a dropped top tube, was a more sporting take on the original design. Some mixte models were made from the same lightweight steel tube sets as men's race bikes and were equipped with the same components.

Although closer to the machines used by their male counterparts, they were still heavier and not suitable for racing. Until WSD arrived, women who wanted to race had to adapt small men's bikes as best they could, altering handlebar width, stem length and seat height to achieve an acceptable position.

This lack of consideration in bike designs did not stop women from taking up the sport. There have been many world-class female athletes and some, regardless of gender, were among the greatest seen in any sport.

BELOW This Peugeot ladies' bike is a good example of the mixte frame with a dropped top tube.

BELOW A loop-framed bike that could be ridden by a rider wearing a skirt or dress. The inset shows a skirt-guard, which prevents any flapping material getting caught in the rear wheel.

Beryl Burton, Great Britain >

Born in Leeds in 1937, Burton never had the stage that her talent warranted because cycling has lagged behind other sports in its inclusion of women. Until 1958 there was no world championship for women and then only two events were introduced. Women's racing wasn't included in the Olympics until 1984 and, incredibly, women only received parity with men in Olympic cycling in 2012. Despite the prejudicial environment, Burton won seven world and an incredible 72 national titles before her death in 1996.

Burton also did something very few women in any sport have done – she broke a men's national record when she rode 446.2km (277¼ miles) in the 1968 British men's 12-hour time-trial championships. Despite the magnitude of her performance, she was not officially awarded the title and neither was her record officially recognized as the rules of the day only recognised male performances at 12 hours. Burton didn't let these outdated ideas affect her and they certainly never diminished her love of cycling.

Jeannie Longo, France >

Jeannie Longo is 59 times a French champion, 13 times a world champion and aged 55 was still competing at the highest level in 2014. She was Olympic road-race champion and silver medallist in the time trial in 1996. In 1992 she won an Olympic silver medal and in the 2000 Games she took bronze. The French champion narrowly missed a medal in the Beijing Olympics (2008), where she finished fourth, again in the time trial.

Marianne Vos, The Netherlands >

Having won world titles in all three types of cycle sport – road, track and off-road in cyclo-cross, something no other man or woman can claim to have achieved – Marianne Vos is thought by many to be the greatest racing cyclist ever. No male athlete has ever matched the breadth of her accomplishments. Some of these titles – cyclo-cross and road-race – she has held simultaneously, another unique achievement in cycling. In the 2008 Olympics, Vos won gold in both road and track events and in 2012 she took the Olympic road-race title in London. As if this list of accomplishments isn't impressive enough, she has also won every women's classic road-race at least once.

It is too early to compare Vos's achievements with those of Jeannie Longo in respect of total victories but, at just 26 years old at the time of writing (2014), she has already beaten the Frenchwoman by winning off-road world titles. (Although Longo did once take a silver medal in the world mountain bike championships.) Comparing Vos with Beryl Burton is impossible because Burton did not have the opportunity to compete for as many world titles or any Olympic titles.

Evolution of the Saddle >

Until very recently, there was no such thing as a race saddle designed for a woman. Even in the construction of standard bicycle saddles, little consideration was given to the anatomical differences between the sexes.

The first bikes had wooden saddles, but not for long. In every country in which bikes were produced, people soon set to work designing something better to sit on. In 1882, John Boultbee Brooks, the son of a horse saddler, was one of the first to come up with something usable. He founded Brooks as a bicycle saddle maker in Birmingham, England in 1866, and the company – Brooks Ltd – still makes stylish leather bicycle saddles today.

In the late 1800s, Brooks led the way in design, releasing a spring version of his saddle soon after the safety bicycle was introduced (see pages 24–5). Although it made the harsh ride more tolerable, many thought the saddle was still a compromise, prioritizing function over comfort. This gave rise to some weird and wonderful inventions, such as two US designs: the Bunker Pneumatic (1892) and Bray's moveable (1898). The former was an air-filled saddle-shaped tube with a hole in the middle. The latter was formed of two lengthways sections that allowed both elements to pivot independently up and down as the rider pedalled.

The saddle explained >

As cycling developed and particularly after racing began, saddles were divided into two main categories: those designed for leisure riding and commuting and those meant for racing.

When we sit down, we do so on part of our pelvis called the ischial tuberosities, commonly called the sit bones. Racing cyclists lean forward, tilting their pelvis and so sit more on the crotch area. In contrast, leisure riders sit more upright, utilizing the traditionally wider, rear part of the saddle. Consequently, radically different saddle shapes evolved to meet these different needs.

Early racing saddles used materials and techniques of the era, so were usually made of leather stretched over a wire frame and riveted in place. Towards the end of the 19th century, springs were introduced to cushion the ride for the leisure cyclist, a method still used today, although gel cushioning has been a popular modern alternative since the 1980s.

Because race saddles are more about support than comfort, most of the padding is removed, both to reduce the energy-absorbing effect of these materials and to minimize weight. In addition, sophisticated materials such as carbon fibre are often used in modern saddles to perform a dual function: that of reducing weight while absorbing some road shock.

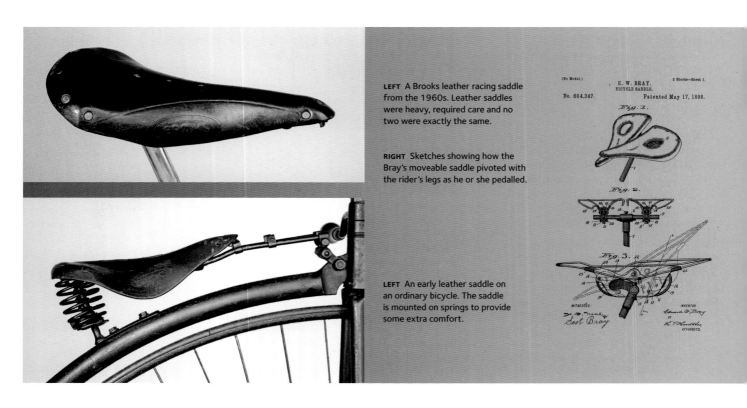

LEFT A Brooks leather racing saddle from the 1960s. Leather saddles were heavy, required care and no two were exactly the same.

RIGHT Sketches showing how the Bray's moveable saddle pivoted with the rider's legs as he or she pedalled.

LEFT An early leather saddle on an ordinary bicycle. The saddle is mounted on springs to provide some extra comfort.

Vive la difference >

Female racing cyclists have been very badly served in the story of the bicycle saddle. Gender-specific saddles for women were first produced in 1992 by Georgina Terry but they were either scaled-down versions of leisure saddles or simply a wider versions of men's racing saddles. Although these had cushioning, what women racers needed was specific support that took account of their anatomy.

Things changed in 2005, when the research and development arm of the Great Britain Cycling Team (the Secret Squirrel Club; see pages 232–49) developed a custom, high-performance racing saddle specifically for women. The saddle was developed in close collaboration with the top female racing cyclists of the day. The design focused almost entirely on the nose section by having a shape that would support the female form in the high G-force environment of track racing. The result was a saddle a third of the length of a standard model and with a stubby nose twice the usual width. There was a grove running down the centre and the whole saddle was fabricated from silicon rubber, custom-drilled on the underside to give precisely the required amount of compliance.

This women's racing saddle was successfully deployed at the Beijing 2008 Olympic Games and provoked an industry rethink about what a women's performance saddle should look like. Now the wide-nose stubby shape has been widely adopted by many saddle manufacturers.

LEFT The modern racing saddle is made from a leather-covered moulded plastic base, sometimes with a thin foam or gel layer sandwiched between base and cover. The whole assembly is mounted on rails. Such saddles are much lighter than all-leather designs. To reduce weight further, some have titanium or even carbon-fibre rails.

Help to choose >

Although choosing a suitable saddle will always be partly a matter of trial and error, two manufacturers – Specialized and Bontrager – have, since 2003, developed a sit-bone measuring device. This uses memory foam, which can guide a cyclist in the direction of a suitable saddle. Other manufacturers recommend a series of anatomical measurements before consulting tables they have prepared for cyclists to pick an appropriate saddle.

Family Cycling >

The saying goes that 'the family that plays together stays together', but I like to think that the family that *cycles* together stays together. Although there is no actual evidence for this, numerous life-long cyclists have been initiated into cycling via a kid's seat, the back of an adapted tandem or from a bicycle sidecar. For many, these early experiences triggered a deep-seated love of the activity.

The first child's seats were mini-saddles that were attached to the top tube of an adult's bike. The child sat astride the seat between the adult rider's arms. Later models incorporated a safety harness but these seats were precarious at best and far from ideal for long journeys. Modern front-carry child's seats, which are also mounted forward of the rider but behind their handlebars, are both safer and more inclusive. They have harnesses and moulded seats for safety and the position allows the child to see where he or she is going.

Bicycle sidecar >
Bicycle sidecars were developed in the 1930s in Germany and their use quickly spread all over Europe. They could carry one or more children and were attached to solo bikes or towards the rear of tandems. The tandem-sidecar combination soon became a favourite of keen cycling families, some of whom used to setup to undertake massive day rides and even extended touring holidays. Bicycle sidecars had a metal tube frame with thin sheet aluminium panels attached. Some were open, but many had the facility to be closed to protect their passengers from the worst of the weather.

RIGHT The whole family can ride together on a triplet. Father, mother and daughter at the back are all pedalling, while one child stands between them and another is inside the sidecar.

The first and simplest child's seats were mini-saddles that were attached to the top tube of an adult's bike. The child sat astride the seat between the adult rider's arms.

Child's seat >

Unlike the modern front-mounted seat, these attach to an adult bike's frame behind the saddle. Early models were made from tubular aluminium with padded wooden seats and safety harnesses. From the 1980s onwards, moulded plastic seats bolted on to aluminium frames became more popular. Each seat has a safety harness and is accredited by road safety authorities of the country they are sold in.

Cycle trailer >

These started out as home-made devices for carrying children but, from the mid 1970s, a number of manufacturers created novel and user-friendly child trailers in which to take children on cycling trips. They range from buggies with one or two seats that have the facility to be covered in inclement weather to attachments with a wheel and pedals to fit on the back of a solo bike, creating an adult–child tandem.

ABOVE A modern child-seat is mounted on a special stand that is attached to the back of the bike. Such seats are made from safety-tested moulded plastic and have security belts for the child.

TOP This child buggy can be towed by almost any kind of bike. It can be detached and pushed, too, and also provide useful storage space.

ABOVE An attachment like this is more than just a trailer; the child can pedal and start to experience longer rides for themselves.

Anatomy of a 1950s Pro Road Race Bike >

Between the first Tour de France in 1903 and the races of the 1950s, bikes evolved slowly. The following two decades saw progress slow even further. Although there were subtle developments and experiments with different materials, bike design hardly changed until the next step forward came in the early 1980s.

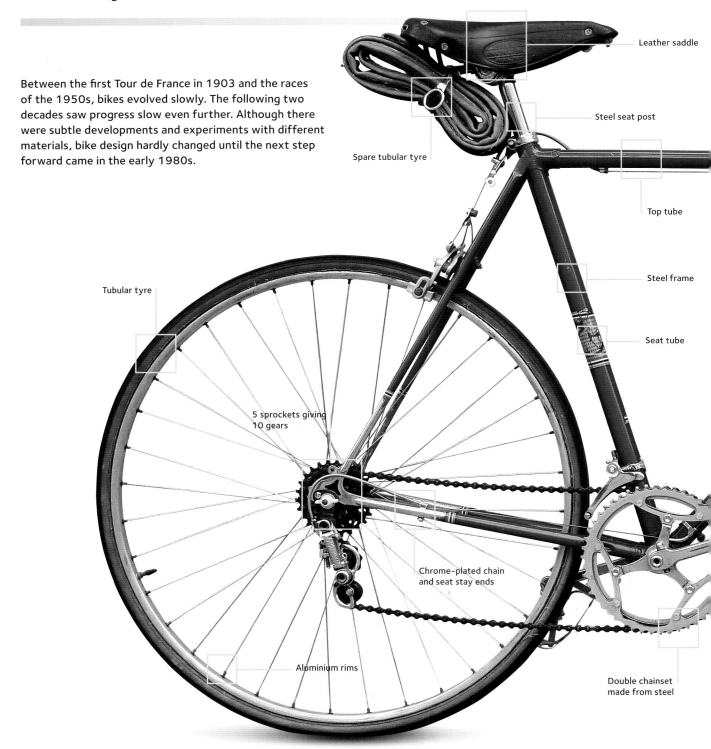

Leather saddle

Steel seat post

Top tube

Steel frame

Seat tube

Spare tubular tyre

Tubular tyre

5 sprockets giving 10 gears

Chrome-plated chain and seat stay ends

Aluminium rims

Double chainset made from steel

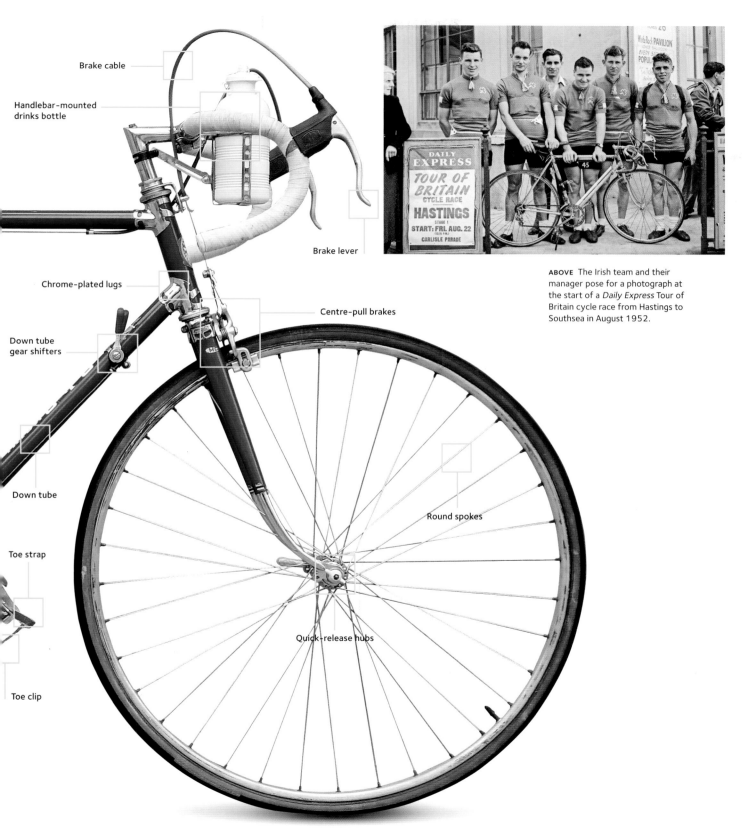

Brake cable

Handlebar-mounted
drinks bottle

Brake lever

Chrome-plated lugs

Centre-pull brakes

Down tube
gear shifters

Down tube

Round spokes

Toe strap

Quick-release hubs

Toe clip

ABOVE The Irish team and their
manager pose for a photograph at
the start of a *Daily Express* Tour of
Britain cycle race from Hastings to
Southsea in August 1952.

Cruising the States >

In the ubiquitous black and white photographs of America in the 1950s, motorcars sported space-age fins and chrome plates. Not surprisingly, many US bikes of the era mirrored this brash, conspicuous automotive styling. In contrast, the cars and bicycles of Europe – perhaps as a result of the hardships suffered in the Second World War – stuck steadfastly to function as their driving design force.

For the next few decades, Americans continued to focus on style, a path that eventually led to the design of the laid back cruiser. Although not very practical, and suitable only for slow riding on city streets, it was an early part of the evolutionary trail that would, ironically, lead to the creation of the most functional bike category of all, the mountain bike.

The Schwinn aerocycle >
Launched in 1934, the first cruiser was a pedal-powered version of Schwinn's B10-E motorbike. Based around the B10-E's cantilevered frame, it sported a false fuel tank between two top tubes and even had integral battery-powered headlights. The massive balloon tyres of the cruiser completed the stylized look.

The Huffy radio bicycle >
The 1950s were the golden era of the cruiser, with companies like Shelby, Monark, Columbia and Roadmaster all producing distinctive models with their own trademark features and detailing. One of the most eye-catching was the Huffy radio bicycle.

Like all cruisers, the Huffy had a long laid-back frame, swept-back handlebars, balloon tyres with white walls and a single gear. But its most distinctive feature was an AM radio built into the void between its twin top tubes. The radio dial can be clearly seen on the side of the bike.

Cruisers were painted in bright colours, using layers of lustrous enamel paint to create striking effects. However, it is arguable that both the Donald Duck bike, which was shaped like the iconic Disney character at the front, and the Cowboy bike, which had tassels hanging all over it, went a step too far.

Clearance of our best selling original Fastback XT101 and Princess models. If you want to save money, order now as quantities are limited on these 1970 models. Others have tried to copy the Fastback, but they never have been able to equal it.
With the 3-speed stick shift you jump away at the start, cruise at high speeds, take steep hills with ease. 15-inch forward, raked steel frame. Chrome-plated high-riser handlebars and fenders. Enamel finish on frame—some in lime, others gold colour ... no choice. Calliper brakes in front and rear. L-bucket dragster saddle of padded black vinyl; big rear safety reflector. Large rear tire (20 x 2.125 inches); with knobby tread. Small front tire (16 x 2 inches). Leg reach adjusts from 22 to 27 inches. Shipped partially assembled.
W61-F 1108—Boy's model. Ord. 69.99. Sale ... **59.99**

The Princess—3-speed girl's bike has all the features of the XT101 above but with gear shift on the handlebars; white saddle. Fuchsia colour. Partially assembled.
W61-F 1107—Ord. 69.99. Sale ... **59.99**

Save $10

on the 1970 model three-speed fastback! Still the leader of the performance pack

59⁹⁹

Design heritage >

Although interest in cruisers declined when Americans saw the race bike as a more efficient way of going cycling, the cruiser's DNA can still be seen in BMX bikes (see pages 154–5 and pages 156–7), in fun bikes such as the Raleigh Chopper (see pages 112–13) and most importantly in mountain bikes (see page 134).

Schwinn Excelsiors, essentially still cruisers, were adapted by the American pioneers for more rigorous terrain. They had unwittingly invented the mountain bike.

Cruisers today >

The cruiser underwent something of a 21st-century rebirth. Americans born in the 1940s and 1950s, who rode the first generation of cruisers as children, snapped up anniversary models produced by the original manufacturers. This encouraged major manufacturers to include one or more cruiser models in their ranges.

The low-rider bike is a modern development on the old cruiser. Young Americans in the 1990s adapted Schwinn StingRay bikes with Easy-Rider handlebars (a nod to the 1969 road movie) and other personal style choices to mimic the look of low-rider motorbikes. These were often custom-painted in exactly the same way motor enthusiasts of the day were customizing their cars.

OPPOSITE The original 1934 Schwinn Streamline Aero Cycle. It wasn't the fastest or most efficient of bikes but, with its luggage carrier and rear-mounted bike stand, it was a thing of beauty and practicality.

LEFT This modern Cruiser, the Felt Burner, is made from light aluminium tubing but has the cantilevered frame design of the original Cruisers, plus a kick-back two-speed hub gear.

Going Clubbing >

The 1920s to the 1950s was the golden era of cycling clubs in Britain. Many people joined clubs as a way to get involved in the growing racing scene, while others participated simply for the social Sunday club runs and never rode competitively. British cycling clubs, at least historically, catered both for leisure cycling as well as racing interests. Cycling clubs in Holland followed the British model, as did those further afield in America and Australia.

Some of the British cycling clubs are very old. The Manchester Wheelers' Club dates back to 1883, when it was known as the Manchester Athletic Bicycle Club. In 1890 it changed its name to the one used today to avoid being confused with the separate Manchester Bicycle Club and the Manchester Athletics Club.

Elsewhere in Europe, cycling clubs were an entirely different entity. There, racing was a stronger driver for club membership, and club racing was often the first step for ambitious young riders hoping to turn professional. European race clubs often arranged travel to events, along with providing other forms of support.

In these countries there were and still are separate bodies that reflected the interests of leisure cyclists and those who enjoyed cycle touring, such as the French capital's Audax Club Parisien, which awarded Brevets Randonneurs certificates to members who covered particular long-distance routes, and the Ordre des Cols Durs, a specialist club for cyclists who enjoyed visiting high places.

Good all-rounders >

In the 1950s, Britain's love affair with the car started in earnest. The increased use of motor vehicles began to erode some of the functions of a club, such as riding to and from races as a body.

Prior to this, bikes were ridden to and from event venues, as well as in the race itself, so they had to be versatile all-rounders; comfortable for long rides yet fast enough to race on. Many chose fixed-wheel bikes with a reversible rear wheel, one side

RIGHT A clubman bike from the 1940s, equipped with everything needed for long days in the saddle. This one was built by the British bike manufacturer RS Gillott.

having a gear ratio suitable for the commute, the other for the race. Owners also added saddlebags to carry their race kit and a change of clothes, as well as mudguards if the weather was wet – all of which were removed before racing.

Heavy-duty wheels were better for the trip to and from the race but were a handicap in competition. This far from ideal situation was solved by the invention of the wheel carrier. A simple bar – often fashioned from a bent spanner – allowed racing wheels to be mounted either side of the bike's forks.

In this fashion, club members rode their fixed-gear bikes, with heavy wheels, mudguards, saddlebags and race wheels, for long distances – often as far as 160km (100 miles) – in order to race. They would stay overnight wherever they could afford (youth hostels were popular). On arrival, they would strip their bikes of unnecessary clutter and change the wheels for competition. The following day, having raced as fast as they could in a time trial of anything between 40km (25 miles) and 160km (100 miles), they'd reassemble their bikes and ride home.

LEFT British club runs typically had an objective, like visiting a historic town. The riders might stop for lunch before setting off home, as it appears the owners of these bikes did back in 1949.

LEFT British club riders might visit a local beauty spot or ancient monument and take all they needed with them for a picnic.

Riding to Work >

There are nine million bicycles in Beijing, or so the song goes. Streets in many other cities around the world ring with the sound of the bicycle bell too. The bike is the most environmentally neutral form of mechanical transport. It does not pollute, it improves the rider's health and takes up less space on crowded roads, making journeys faster and more efficient. It is the ideal solution for short journeys.

The European cycling infrastructure >
The short-sighted planning of many European countries, including Britain, means that they now face the difficult job of integrating a cycle network into road systems that were originally designed to prioritize motor vehicles. One exception is Belgium, whose town and rural road planners incorporated bikes into national transport infrastructures almost as soon as the bicycle was invented.

The Netherlands has also been held up as a transport success story, but space was only made for bikes there following a major re-think of transport policy resulting from an increase in bicycle use at the time of the oil crisis in the early 1970s. Nowadays, 70 per cent of all journeys in Holland under 7.5km (4 miles) are made by bike and 50 per cent of school children ride to school.

All over Europe, the bike is being revived as a modern transport solution. To facilitate this, special carriageways, such as shared bus and bike lanes in Mannheim, Germany, are being built. In other cities where space doesn't allow for segregated cycling infrastructure, innovative methods have been used, such as throughways that are out-of-bounds to cars at certain times of day. Paris is one of the cities also working hard to re-popularize cycling as a mode of transport, with 440 km (270 miles) of cycle paths across the city, while Copenhagen is one of the most cycling-friendly cities in the world, with an extensive and well-designed system of cycle tracks.

Cycling infrastructure around the rest of the world >
Western Australia's Ride2Work initiative encourages people to commute by bike by providing practical support and advice, and via their annual Ride2Work day. It also campaigns for employers to provide bike storage facilities at the workplace. It's a start but Australia's road network in general does not cater to cyclists.

Large areas of the United States have a similarly fragmented approach to cycling and like Australia, there are notable exceptions. Davis, in California, has the highest percentage of American bicycle commutes at 19.1 per cent of journeys. This contrasts sharply with Bloomington, Indiana, where only 3.9 per cent of journeys are made on bikes. These statistics show the need for national strategies that make cycling look and feel safe. Only then will cycling be a real choice for people.

RIGHT Hire bikes at a bike station in Hangzhou, China. The city's bike-sharing programme is the world's largest, with more than 65,000 bikes and 2,500 bike stations. The scheme is integrated with other forms of public transport, with bike stations being located near bus and water-taxi stops.

BOTTOM RIGHT A city bike-sharing station in Milan, Italy. With thousands of bicycles and more than 200 bike stations, BikeMi is one of the most successful bike-sharing systems in the world.

FAR RIGHT The Vélib' scheme in Paris, which has been running since 2007, has 20,000 bikes available 24/7 from 1,800 docking stations.

Bicycle commuting >

More people are choosing cycling as their preferred means of transport, although the trend is slow (only 2 per cent of journeys in the UK) and tends to be localized to big cities. However, you have only to look at the streets of London for an example of how many more cyclists there are today compared to 10 years ago.

In Europe as a whole the picture is better. Data from seven European countries shows that 3–28 per cent of all trips are now made by bike, with the Netherlands leading the way at 34 per cent.

Governments worldwide are now making efforts to facilitate this trend. Some trains in Europe and North America now have on-board bike racks and bike carriages, so commuters (and leisure users) can cycle at each end of their journey. Folding bikes (see pages 114–15), are popular with commuters. Yet if cycling is to be a viable choice for people travelling to work, then the end-of-journey experience also needs to be addressed. Secure, convenient bike-racks and access to showers is essential. Palo Alto, California, has introduced not only secure bike parking at rail stations but showers and changing facilities too, called Bikestations.

Sharing bicycles >

Perhaps the biggest thing Holland has contributed to global cycle commuting is the template for a bicycle-sharing system – the White Bicycle Plan. However, the scheme was far from a success. The story goes that, in 1965, the radical, anti-establishment group Provo painted 50 old bikes white and left them on the streets of Amsterdam, with the understanding that they could be freely used and then left for somebody else. However, within a month of their release, most of the bikes had either been stolen, or impounded by police, who cited a city bylaw that forbade anyone leaving an unlocked, unattended bike.

Provo's Luud Schimmelpennink has since played down the 1960s experiment, explaining that no more than about 10 bikes were put out 'as a suggestion of the bigger idea'. Nevertheless, the scheme planted a seed and was copied on a larger scale in cities such as Helsinki, Copenhagen and Lyon.

Aided by smart card technology, bike-share schemes have blossomed worldwide. Stockholm City Bikes are the Swedish capital's equivalent of Provo. BikeMi runs in Milan and has an average of 5,000 users per day, while Velib, the Parisian equivalent, is the second most extensive in the world. Barclays Cycle Hire, whose bikes are more commonly known as Boris Bikes after the Mayor of London, is that city's bicycle sharing scheme.

China's Hangzhou Public Bicycle Programme has 65,500 bikes operating from 2,500 docking stations. It is one of 19 bicycle sharing schemes in China. Started in 2008 as a response to concerns about traffic congestion it's been a huge success. Bikes conveniently integrate with the transport system in Hangzhou. The scheme is – or could be – a blueprint for integrated transport plans throughout the world.

ABOVE LEFT Savvy bicycle commuters crossing Waterloo Bridge, London, in April 2014, unaffected by a Tube strike going on at the time.

RIGHT Beatle John Lennon and artist Yoko Ono trying out a bicycle that they received as part of Provo's White Bike Plan in 1969 during their honeymoon 'bed-in' in the Hilton Hotel in Amsterdam.

BELOW Boris Bikes lined up ready to use in London. Members and casual users can unlock bikes from docking stations all over the city. The hirer then places the bike in another station at or near their destination. The bikes are highly size adjustable, have hub brakes, three-speed gears and are very durable.

Chapter 7
The Age of Innovation >

Models from the Catherine Harle
Modelling Agency ride in single file
along a country road in 1967.

The Mini and the Moulton >

The 1960s witnessed a revolution in politics, music, style and design. The world finally shook off the effects of the Second World War and many people, especially the young, began to question the old order. The African-American Civil Rights Movement campaigned for an end to racial segregation and discrimination, and the end of army conscription in Britain coincided with the emergence of a youth culture. Rock 'n' roll music from America and British groups The Beatles and The Rolling Stones influenced lifestyle, fashion, attitudes and language. This cultural explosion spread throughout Europe and the rest of the world. Almost everything that had gone before was challenged. This mood was reflected in sixties design. Nothing was left untouched – not even the bicycle.

Although Dr Alex Moulton is the name that most people associate with the small-wheeled bikes, he was not the first to experiment with wheel size. Back in 1911, Frenchman Paul de Vivie – who in 1905 developed one of the first derailleur gears (see page 45) – talked of his trials with wheel sizes as small as 20-in but, despite the potential, the concept did not capture people's imagination.

In 1962, 50 years after de Vivie's original experiments, Moulton introduced his refined small wheeler to a public already in the throes of a fashion revolution and primed for avant-garde ideas. His creation received a much warmer reception.

The Moulton bike of 1962 was heavily influenced by the Mini motorcar (launched in 1959) and the Italian Vespa scooter design of the 1950s. The Moulton drove bike design along a new path. It was the first commercially successful departure from diamond-frame design (see pages 24–5), the configuration that had held sway since the 1890s. Its small wheels with narrow, high-pressure tyres, coupled with effective suspension, created a nippy, modern-looking form of transport that would prove hugely popular.

Moulton was not alone in his belief that small wheels were the future. Reise & Müller in Germany also used small wheels and a new way of looking at the bike frame in its Birdy folding bikes.

RIGHT The original Moulton bike was stylish and wonderfully adaptable, making it perfect for short journeys that might otherwise have been made by car.

Dr Alex Moulton >

Born in Stratford-on-Avon in 1920, Alex Moulton was educated first at Marlborough College, then Cambridge University, where he graduated as an engineer and worked for the Bristol Aeroplane Company. He joined the family rubber business after the Second World War and pioneered research into the use of rubber in vehicle suspension systems. Moulton both invented and developed the hydroelastic suspension system for the Mini motorcar, a system that replaced leaf springs with rubber cones.

Moulton was involved with the motor industry for most of his life and only turned his engineering mind to bikes in the late 1950s. When he did, Moulton fell in love with cycling. In 1966, Moulton said this about his passion: 'Having spent most of my life involved in automotive engineering, I find it interesting to reflect that the automobile, with all its convenience of use and fascination to the engineer, was an inevitable evolution from the horse-drawn chariot; whereas the bicycle, with its human propulsion and single track, is an extraordinary and unlikely device to have been created. The contrast in the activity of motoring and cycling could not be more profound. The driver, who sits in a low, enclosed, air-conditioned environment or "cage", is only sensually and nervously involved and is isolated from the surroundings; whereas the cyclist, with a high sightline and in the open air, is physically involved, and proceeds in that miraculous way entirely by his or her own effort of health-giving exercise, with nervous relaxation and spiritual uplift even.'

RIGHT Dr Alex Moulton loved all machines with two wheels and the feeling they gave the rider of being out in the open.

LEFT/BELOW The Birdy folding bike from Germany is perfect for commuting. It has full suspension, small wheels and folds up into a shape that is easy to carry and store.

Wheel Size Theory >

700c, 650b, 26in. No, this isn't a code; these numbers are wheel sizes – a measure of the wheel's diameter including its tyre. They are all wheels sizes available today on adult bikes. With each comes advantages and disadvantages.

The unit used depends on where that sized wheel was first used: the US and UK for imperial, Europe for metric. Rather than try to understand why the cycling world persists in flitting between imperial and metric when talking about wheel size, it's probably best to recognize and accept it for what it is – a benign quirk of the global two-wheeled community. The vast majority of wheels on road and cyclo-cross bikes are 700c (see pages 152–3). Some road bikes for people 1.57m (5ft 2in) or shorter come with 26-in or 650-c wheels. Although smaller wheels on road bikes present challenges with gear ratios, and many tyres are only are available in 700c, smaller wheels help keep all geometry in proportion and can give a better ride experience for the more compact rider.

Time-trial bikes used to have a 26-in or even a 24-in front wheel, because it was thought to be more aerodynamic than a 700c. The aerodynamic argument for small wheels – that wheels cause turbulence when they rotate, so smaller wheels must mean less turbulence – is not convincing. The reduced amount of wheel must be replaced by frame tubes in order to bridge the gap between rider and bike. The discussion also does not factor in the extra ground friction that small wheels create.

Newton's law >

Whether the designers were right or not about wheel sizes on time-trial bikes became a moot point when cycling's governing body – the Union Cyclist International (UCI) – decided both wheels on any race bike had to be the same diameter.

Swiss pro riders, Tony Rominger and Alex Zülle, along with French star, Laurent Jalabert, experimented in the 1990s with 26-in wheels on road bikes. They chose the wheels because of an engineering premise based on Newton's second law of motion around rotational mass. When a bike travels, the wheels rotate and more force is required to accelerate the mass of the wheel than that of the frame, due to rotational inertia. However, this is only a factor while under acceleration. At constant speeds, aerodynamics are much bigger factor. Again, increased ground friction, as well as poorer handling characteristics, do not seem to have been considered.

RIGHT Swiss pro-rider Alex Zülle was one of a few racers who experimented with 26-in wheels in road races during the 1990s. Here he is riding a mountain time-trial stage in the 1994 Tour de France.

Mountain bikes >

The standard wheel size for an adult mountain bike has historically been 26 inches. It was the size the first American mountain bike producers chose, presumably because they thought 26-in wheels worked best with fat, knobbly tyres for rolling over rough ground.

Larger wheels were first experimented with by English off-road cycling pioneer Geoff Apps in collaboration with MTB legend Gary Fisher in 1981. But it wasn't until the turn of the 21st century that manufacturers took alternatives to the trusty 26-in seriously. The 29-in wheel (the approximate diameter when a fat mountain bike tyre is included) rolls over obstacles and handles bumps better than bikes that are fitted with 26-in wheels. They first appeared on Wilderness Trail Bikes in 1999. However, they are thought by experts to be less agile than their smaller brethren and, due to the extra rotational weight, do not accelerate as quickly. For most users, the benefits outweigh the downsides and 29ers – as the bikes are now commonly known – quickly became the norm in North America and in much of Europe. For all but the biggest riders wanting to use full-suspension bikes, 29-in wheels presented significant technical challenges for the manufacturers. The larger wheels leave little room on the frame for suspension to function.

Around 2011, an industry consensus seemed to have been reached and the 650b or 27.5-in wheel emerged as the accepted norm for full suspension bikes. (Or at least one of them – amongst off-road aficionados the debate over the ideal wheel size rages on.) This halfway-house solution rolls better than the traditional 26-in wheel, is less ponderous than the 29er and leaves enough room around the frame to fit in meaningful rear suspension.

LEFT The universal choice of wheel size for mountain bikes was originally 26-in but manufacturers are now trying different sizes, with 29ers and 27.5s proving popular.

The Raleigh Chopper >

For kids in the 1970s, the Raleigh Chopper was the pinnacle of transport fashion. Although Schwinn's StingRay is thought by many to be the first in the genre (an engineless copy of Harley-Davidson-style motorbike) the biggest seller on both sides of the Atlantic was the Chopper.

There is some dispute over who actually designed Raleigh's iconic machine. Tom Karen of British design company Ogle claims to this day that he invented the Chopper. It's an assertion that has always been disputed by Raleigh, which sites Alan Oakley, who led their in-house design team, as the true father of the classic 70s bike.

The bike was trialled in North America during the late 1960s and introduced to the British market in 1970. Available in single, three- or five-speed versions, the Chopper had many motorbike-inspired design features. It cost £35 for the basic model and £55 for the deluxe.

On the Mark 2 version, launched in 1972, the stick gear shift was replaced by a T-bar, mimicking the gear selector on automatic cars of the era. The Mark 2's handlebars and stem were welded to form a single unit, which prevented kids customizing their bikes by angling the handlebars back and making them less safe.

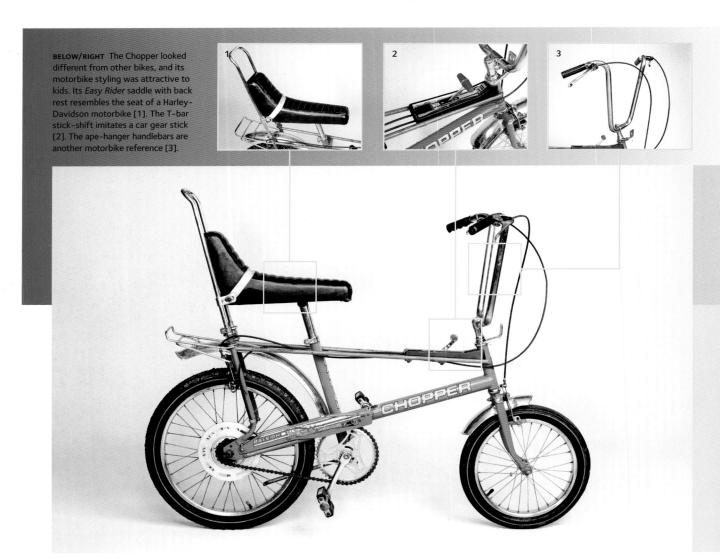

BELOW/RIGHT The Chopper looked different from other bikes, and its motorbike styling was attractive to kids. Its *Easy Rider* saddle with back rest resembles the seat of a Harley-Davidson motorbike [1]. The T-bar stick-shift imitates a car gear stick [2]. The ape-hanger handlebars are another motorbike reference [3].

Although it is looked on fondly now, there was criticism of the Chopper even at the height of its popularity. With the bench-style saddle putting the centre of gravity over the rear wheel, it was tricky to ride and could easily 'wheelie' unintentionally.

Although it is looked on fondly now, there was criticism of the Chopper even at the height of its popularity. With the bench-style saddle putting the centre of gravity over the rear wheel, it was tricky to ride and could easily 'wheelie' unintentionally. The long, motorbike-like seat allowed the rider to carry a passenger. The top tube-mounted gear stick was also contentious — it was the cause of some nasty injuries. Production of The Chopper stopped in 1981, by which time the bicycle motocross (BMX) craze (see pages 154–7) was in full swing.

ABOVE/LEFT Raleigh's advertising campaign for the Chopper made the bike look cool, as well as referencing other original British inventions, such as the hovercraft.

The Chopper returns >
Riding a wave of nostalgia, Raleigh reintroduced the Chopper with a Mark 3 version in 2004. The precariously positioned gear stick had gone from the top tube but was remembered with a graphics sticker. The bench saddle was replaced with a more conventional model to comply with modern safety standards.

Folding Bikes >

The convenience of a bike that can be easily folded when not in use indelibly changed the way commuters travelled.

Dahon >

Folding bikes were first used to allow army infantrymen to move quickly over suitable terrain. Later, the introduction of small wheels allowed them to fold into an even more compact and carryable size, which was key to their widespread adoption by commuters.

It was the oil crisis of the 1970s that inspired Dr David Hon, an American physicist turned entrepreneur, to develop a folding bike. The Hon Convertible was launched in 1982 and was a huge success. In 1995, Hon opened the Dahon factory in Shenzhen, China, which now produces around half a million folding bikes every year.

In 2009, Dr Hon made the following comments about folding bikes: 'In the last 25 years we have seen first-hand the varied ways that portable bikes can contribute to a greener and better way of life. City governments and transport authorities have eased regulations and opened new opportunities for commuters to use bikes as an intermodal means of personal mobility. These are very promising first steps into an even more widespread rethinking of transportation in the next few years.'

Brompton >

To Brits, Brompton is to folding bikes what Hoover is to vacuum cleaners. The name is so synonymous that many people refer to all folding bikes as Bromptons. The brand was conceived in 1970 by Andrew Ritchie in a room overlooking London's Brompton Oratory.

The modular design is a true classic and has remained fundamentally unchanged since the original patent was filed by Ritchie in 1979. It folds easily, is light to carry and quick to stow. At the other end of a journey – or after a day at work – reassembly takes only seconds. On the road, the Brompton is both nippy and comfortable.

Other companies produced folding bikes before Brompton came on the scene but something about the London-designed bike caught people's imagination. Production began and, with a few bumps along the way, Brompton evolved, first into a brand and then into a movement.

LEFT An example of a very racy-looking Dahon, with deep-section aerodynamic wheels and a Shimano STI gear system.

A Brompton bike ready to ride [1], and the same bike folded for carrying or storage [2].

Folding racers >

In 2006, the Spanish importer of Brompton bikes started the Brompton World Championships. For this, contestants can wear shorts and cycling shoes but they must also have on a shirt, tie and jacket. The first two championship races were held in Barcelona, then they switched to Blenheim Palace, in the UK. Competition has been fierce, with triple Tour of Spain winner Roberto Heras winning the 2009 edition.

Good fun and hard fought as Brompton races are, the bikes are not designed for racing. There is a company, however, whose bikes do exactly that. Airnimal bikes are race quality folding bikes and are sold all over the world. The company produce mountain, road and even time-trial versions that, post-race, can be folded up and placed into a standard-sized suitcase.

Torque couplers >

A torque coupling device can be fitted to any bike, turning them into a folder. As the top tube and down tube have to be cut and shortened slightly, torque couplings must be fitted by specialists. Once in place, the couplings allow the bike to be assembled and disassembled whenever the owner wishes.

ABOVE Torque coupling. The tube on the right fits into the one on the left, then the left tube-sleeve screws down onto the right tube-thread, locking the two together.

TOP Action from the Brompton World Championships in 2013. Riders must assemble their bikes from folded and race in an everyday shirt and jacket.

ABOVE An Airnimal folding mountain bike, with a specification that means it is good enough for racing.

Plastic Fantastic >

Since the bicycle was first conceived, manufacturers have experimented with every construction material imaginable, including plastic. Most persistent with this material was the French component manufacturing giant Simplex which, in the 1960s, produced a plastic bicycle using cutting-edge processes.

The company was founded by Lucien Charles Hyppolyte Juy, a French bike shop owner. In 1928, he developed an improved version of Campagnolo's early derailleur (see pages 52–3). With an arm and sprung pulley wheel in a cage, it resembled a modern-day derailleur. Shifts were made by the pulley wheel being pushed and pulled sideways by a cable-operated push rod. The derailleur differed from Campagnolo's early derailleur in that the jockey wheels were part of the derailleur mechanism, as they are today.

Juy continued developing his front and rear derailleurs, both of which were regularly used by pro riders in races such as the

Tour de France. The Simplex products proved very popular and only when Tullio Campagnolo introduced the first parallelogram derailleur in the early 1950s did the Italian regain the upper hand.

Campagnolo spurred Juy on and by 1961 he had developed a parallelogram derailleur of his own, the Export 61. Again, many thought the Simplex version was superior to Campagnolo's and it sold well, but the Export 61 was an improvement rather than a unique design. It wasn't a situation Juy – who regarded himself as a revolutionary – was content to accept.

Not satisfied merely to follow, Juy, who had started searching for alternative materials to aluminium, came across polyoxymethylene, a thermoplastic that had been christened 'Delrin' by DuPont, who had developed it. It was light, strong and corrosion-proof, and Juy thought it had real potential as a alternative to metal. DuPont may have been the first to see the value of the hard plastic as an engineering solution but it was Hermann Staudinger who was its true inventor. In fact, the German chemist received the 1953 Nobel Prize for developing the thermoplastic.

Juy used Delrin to produce the first plastic derailleur – the Prestige 532 – in 1962. The whole body of both the front and

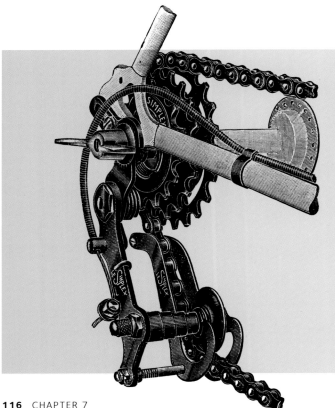

LEFT The Champion du Monde, an early Simplex gear, designed by Lucien Juy. The sprung cage moves the chain from larger to smaller sprockets when pulled by the control cable. When the cable releases, the cage springs back, shifting the chain from smaller to larger sprocket.

ABOVE A Simplex Prestige rear mechanism. The black parts of the derailleur body are made from the engineering thermoplastic Delrin, developed by DuPont.

The bikes, which were delivered flat-packed with tools for home assembly, suffered from broken parts and packs without all the required components. The adverse publicity had a devastating effect on the already beleaguered enterprise and the company ceased production in 1985.

rear derailleurs was made from the new material. Through the 1960s, Juy continued to refine the design and several Delrin iterations followed. Simplex also made plastic shifters to match their gear system but switched back to aluminium in the 1970s, retaining plastic only for the cheaper end of its range.

Plastic bikes >

In 1981, Swedish company, Itera, was the first to present a commercially available plastic bike to the public and press. The launch attracted a huge amount of publicity and high demand was anticipated but sadly it never materialized. In addition, the bikes, which were delivered flat-packed with tools for home assembly, suffered from broken parts and packs without all the required components. The adverse publicity had a devastating effect on the already beleaguered enterprise and the company ceased production in 1985. Around 30,000 bikes were produced. The remainder of the stock was sold to enterprises in the Caribbean where it was thought the lack of metal would be advantageous in the humid environment. Some Itera bikes are still being ridden in the West Indies today.

Although it was the first to be commercially available, Itera's bike may not have been the first to be made from plastic. In the 1970s, a group of Americans tried to produce an injection-moulded bike but failed to get beyond the prototype phase. Today, the aerospace giant, Airbus, is still researching materials and methods to make a commercially viable plastic bike.

BELOW Even the wheels and spokes of the Itera bike were made of plastic, but the end result was nevertheless a heavy and quite clumsy bicycle.

Artisan Frame Builders >

Today, bikes are manufactured rather than built. For me, the distinction is determined by where and how the human being is involved in the creation process. The love, expertise and skill used to make a modern bike comes mostly at the design stage, after which machines take over and implement the makers' plans. With older bikes, human expertise was required for every stage of production; from design, construction, painting, building wheels, and even to assembly, where master mechanics adjusted components to add the last percentage points of performance.

There are still craftsmen and women practising bike production today. This small group of mechanical artists can be found tucked away in specialist bike shops and working with the pro cycle teams. The creation of a hand-built frame requires a combination of art, science and skill. The result is usually a thing of functional beauty.

One of the hand-built frame's greatest modern exponents is the Italian legend, Dario Pegoretti. At his base in Trentino in the Dolomite mountains, he continues to craft individual masterpieces for discerning customers. His work is held in such high regard that there is a lengthy waiting list for a Pegoretti

machine. Many successful racers, who spend their working lives on carbon fibre, have bought one of his steel bikes. Pegoretti even paints the frames he makes himself, each one unique.

Steel tubes, the traditional medium for frame builders, can be joined together either by welding or brazing. Brazing is a low-temperature process during which the brazing medium, which is often silver-based, is melted and drawn into the tube joints by capillary action. Brazed joints are created by building up the material into joints called fillets, or by connecting the tubes with steel junctions called lugs.

Hand-building with aluminium and titanium is much more challenging. As it is liable to catch fire at high temperature, titanium is particularly difficult to work with and can only be welded in an inert gas.

Made to measure >
The process starts by measuring the client. From there, the builder then draws the frame and cuts the tubes to the required length. These are then mitred at the ends, to ensure a perfect match when the custom-length tubes are placed together for brazing.

The tubes are then clamped to the jig for welding or brazing. Once the tubes are joined, the frame has all its burs and extra

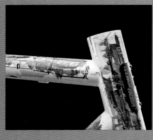

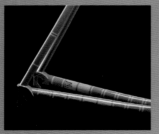

RIGHT Dario Pegoretti is as much an artist as he is a master frame builder. No two of his Responsorium frames have the same paint finish.

'It is my hope that the frames I make are used on the roads and not hung as art on the wall.'

DARIO PEGORETTI, ITALY

The love, expertise and skill used to make a modern bike comes mostly at the design stage, after which machines take over and implement the makers' plans. With older bikes, human expertise was required for every stage of production.

material removed. It's then cleaned – often by sand blasting – in readiness for painting, plating or both.

Some artisan frame builders still finish their frames by hand, building up layers of enamel paint to create custom designs with a lustrous finish.

BELOW Artisan bike producers like Vanilla Workshop aren't constrained by appealing to mass-market trends, so they can let their imaginations run wild. This single-speed bike is not only elegant but practical, with fat tyres for comfort and cantilever brakes for stopping power.

RIGHT This bicycle in vivid red was made in the Vanilla Workshop, a collaboration of frame builders and other craftspeople based in Portland, Oregon, who build custom-make bikes of the highest quality.

'I am inspired by French bikes of the 1950s and 1960s, but I do not make reproductions of them. I try to honour the emotional and visual spirit of those bikes and add my own elements and style.'

PETER WEIGLE, UNITED STATES

RIGHT Peter Weigle takes his inspiration from 1950s and 1960s French bikes. He matches it with modern construction techniques and a sharp eye for detail. The integrated rear lamp bracket on this bike is a thing of simple beauty.

'The experience is like going to a tailor for a bespoke fitting rather than picking something from a rail on the high street. Make an appointment and come for a chat at our studio. We will discuss your project in detail, take your measurements and after ten weeks you can collect your dream bike.'

ANTONIO TAVERNA, ITALY

Road Bike Brakes – The Early Years >

You could be forgiven for thinking that this is one part of the bicycle's story that racing did not drive. Surely, racers do not want to slow down. But you would be wrong. Just like motor racers, bike riders need precision stopping power – brakes that are predictable, dependable, durable and preferably, lightweight.

In the beginning >

Although racing encouraged brake development, safety was, unsurprisingly, the original driving factor for fitting them to the bicycle. The spoon brake was the first to appear on a bicycle, a simple device activated from the handlebars by a lever or chord. The spoon brake pushed down on the front wheel creating friction and retarding speed. It was not very powerful.

Most braking on early bikes, which were all direct-drive with the pedals attached to the front wheel (see pages 12–13), was done by the rider's legs pushing back on the pedals. When riders wanted more speed and front wheels got bigger, applying a spoon brake too vigorously could actually be dangerous. The rider's own body mass, perched almost over the wheel's centre of gravity, could easily fall forward – known as taking a header – if speed dropped too quickly.

The answer to this was the first calliper brake, which used rubber brake blocks acting on the rim of the smaller rear wheel to slow the rider. Spoon brakes were still used on the first safety bicycles (see pages 24–5) until pneumatic tyres were developed (see pages 26–7). Braking on the tyre wore its tread down quickly, so another solution was needed.

The man with the answer was Abram W Duck, of Duck's Cyclery in Oakland, California. He used lever-activated twin rubber rollers to work on the front tyre and went into production with his Duck brake in 1897.

The freewheel (see page 44) was developed shortly after that. It made cycling on downhill slopes and corners safer because it allowed the rider to stop pedalling. The ability to freewheel removed the danger of the pedal striking the ground when the bike leaned over as pedals could be held level. However, freewheels took away the rider's ability to slow the bike by resisting the revolution of the pedals, which made brake development an even higher priority.

The coaster brake – a drum brake inside the hub, applied by back-pedalling – was invented in 1898. For the first part of the 20th century, coaster brakes were the only type of brake found on most American bikes.

RIGHT The rod- and lever-operated spoon brakes that were used on ordinary bikes were elegant and simple but had to be applied very carefully if the rider were to avoid taking a header (see page 18).

Meanwhile, in Europe racing had taken over bike brake design. The first racers used simple spoon brakes but soon adopted cable-actuated calliper brakes. Early brakes of this type were made from two steel spring-loaded callipers with a central pivot bolt fixing them to the bike's frame. They were joined by cable to a handlebar-mounted lever. Pull the lever and the callipers moved inwards, bringing two rubber brake blocks into contact with the wheel rim. The principle of this system – called a rim brake – is the basis of most brakes in use today.

BELOW This Raleigh ladies' bike is equipped with rod brakes, which are simple but effective and very easy to maintain.

The Flying Pigeon >

The Tianjin Flying Pigeon Cycle Manufacture Co, from China, has produced more than 500 million bikes, making it the biggest vehicle manufacturer in the world. Currently, the company makes several models of bikes, including an electric one, but its best-selling are three standard models: the PA02 and PA06 (both men's bikes) and the PB13 (for women). They are all single-speed, painted black and made from steel. They are all fitted with crude but reliable rod brakes.

Rod brakes are a natural extension of the rod-and-lever-actuated spoon brake. Instead of cables, a series of steel rods run from brake levers to brake pads through a series of linkages. When the handlebar-mounted lever is pulled, the rods draw the brake pads up from beneath the wheel rim. Rims have to be a special shape to work with rod brakes; this shape is called the Westwood profile after Frederick Westwood, who invented them.

ABOVE/LEFT Rod brakes work so well that they are used on Flying Pigeon bikes made in China today. The inset shows the Westwood wheel rim, which has a unique cross-section suited specifically to rod brakes.

Road Bike Brakes – Getting Up to Date >

Although safety requirements triggered the invention of bicycle brakes, the demands of racing soon took over. Racers need brakes they can rely on, that respond to their input not only in time but also in force of application. Thus, racers talk about brake modulation or 'feel'. What they are describing is how the brake responds to input: a slight pull on the brake levers should give a correspondingly light braking force. Equally, when the pull is harder, there should be a commensurate response. The relationship between pull and application should progress smoothly, with no snatching at the wheel rim, which can cause a dangerous wheel lock-up.

The first steel callipers that racers used were not very effective. Manufacturers had made these of thin steel plate – to keep weight down – which flexed under hard braking. Aluminium was a much better option. Callipers could be made thicker, to reduce flex, while still remaining light.

Balilla brakes, produced in Italy from the 1940s, were a market leader. Unlike a lot of cycling equipment, they were not named after their inventor but after an Italian folk hero. In 1746, when the Austrians occupied Genoa, a young boy called Balilla

was said to have thrown a pebble at an Austrian official. His actions started a revolt that resulted in the withdrawal of the Austrian forces from Genoa and the eventual victory of the Genoese people. A statue of Balilla throwing the pebble can be found in the middle of Genoa.

Balilla the company was a big player in the move from side-pull to centre-pull brakes. Side-pull callipers had a major disadvantage – a tendency to one sidedness when applied. One brake block would contact the rim before the other, reducing the overall effectiveness. The immediate answer was the centre-pull brake.

Originally developed in the 1950s, centre-pull brakes came into wider use in the early 1960s. Leading the way were Balilla (with its Tipo Corsa 61 brake), Italian company Universal and Manufacture Arvernoise de Freins et Accessoires pour Cycles (MAFAC) of France.

Centre-pulls were operated via a central cable, ensuring each brake pad contacted the rim at the same time and with equal force. Centre-pulls enabled racers to brake later and to modulate their braking more accurately. Those using centre-pull brakes enjoyed a competitive advantage over those with early calliper systems. The only drawback of the new system was the extra material needed, making centre-pull brakes heavier than side-pulls.

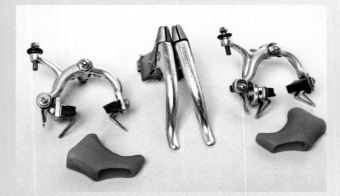

RIGHT, TOP AND BOTTOM
Campagnolo's original Record brakes, one of the best single-pivot brakes ever made, are pictured above a later pair of aerodynamic Delta brakes from the same company.

FAR RIGHT, TOP AND BOTTOM
Mafac Racer centre-pull brakes were very effective, very adjustable but big and heavy. Pictured below is a Universal side-pull brake, which was considered the best available until Campagnolo brought out the Record brake.

Side-pull solution >

In 1951, Universal produced a side-pull calliper with a very effective quick-release mechanism, allowing the pads to be set much closer to the rim, giving more braking power and control. Thanks to the quick release, the pads could be easily backed off from the rim, allowing a fast wheel change in a race situation.

Campagnolo introduced its Record brake in 1968. It was an engineering masterpiece that functioned as well as a single-pivot side-pull brake could work. And in those words lies the answer to the side-pull's inherent weakness – its single pivot. Even the best single-pivot callipers tended to pull to one side over repeated actuations and required frequent re-centring. The dual-pivot side-pull, first developed by Shimano in 1990, cured that by transferring the pull exerted by the rider's hands on the brake lever equally between both sides of the calliper. The new dual-pivot design was also a lot more powerful than even the best of the single pivot brakes.

Shimano actually had the solution to the single-pivot brake in 1984 but had been producing so many innovative designs, it had been forgotten in their design department until 1990!

But even before then, Shimano, with its seemingly endless stream of inventive solutions, worried Campagnolo.

In 1984, the Italian company launched the radical-looking Delta brake. The futuristic triangular-shaped cover (from which the brake got its name) housed a complicated, parallelogram mechanism that gave powerful braking but was a heavy solution to a problem that had already been solved. The Delta was also difficult to set up and it soon fell out of favour.

Shimano's success with their dual-pivot design in 1990 was a step change in quality and efficiency for road brake bikes. Campagnolo quickly developed a dual-pivot design of its own. The majority of today's road bikes are now are fitted with dual-pivot, side-pull brakes.

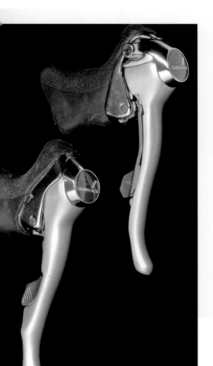

LEFT A Shimano Dura Ace STI. This was the pioneer system that combined brake levers and gear shifters in one unit.

RIGHT Another pioneer, this time Shimano's dual-pivot callipers. Dual pivot provided equal force on either side of the wheel rim and gave better modulation when the brakes were applied.

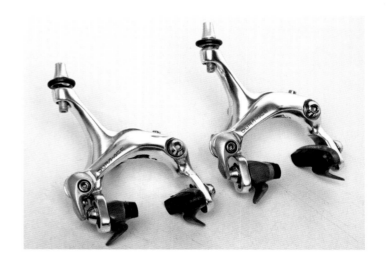

The Group-Set >

In the late 1960s, parts manufacturers who had previously specialized in making one kind of bicycle component began to expand their ranges of products and offer several pieces of equipment as a package – known as a 'group-set'. Today, this term is understood to refer to a combination of brakes, gears, drive-train and, sometimes, seat post.

Before the emergence of group-sets, bikes would carry equipment from several different manufacturers, many of which specialized in just one type of component. For example, Italian companies Balilla and Universal produced only brakes, whereas the French company Spécialités TA specialized in chain-sets and chain-rings, although it also made a few other items. So it was common for a bike to have Balilla brakes and a Spécialités TA chainset.

The Peugeot team bikes of the 1960s are a good example of this phenomenon. They carried gears, seat posts and hubs made by Simplex (France), wheel rims by Mavic (France), chain-sets from Stronglight (France), brakes from Manufacture Arvernoise de Freins et Accessoires pour Cycles (France), pedals from Campagnolo (Italy) and saddles from Brooks (Britain).

Birth of the group set >

By the 1960s the Italian manufacturing company Campagnolo had developed its front and rear mechanisms, hubs, chain-sets, bottom brackets, pedals and seat posts to a point of near perfection for the times. When Campagnolo finally added brakes to their range of products in 1969, they began to offer all of these components for sale as a single product: The Nuovo Record group-set. At the time, Campagnolo did not make a freewheel or chain; instead, they recommended using those from another Italian company, Regina. But the invention of the free-hub changed things even more.

The Japanese manufacturer Suntour developed a free-hub in 1969. It was the forerunner of the standard free-hub that is found today, in which the freewheel mechanism is attached to the rear hub and the sprockets are mounted on it. The multi-sprocket freewheels that Regina made were one-piece, sprocket-and-freewheel units that screwed on to the hub. Complete group-sets including sprockets and chains did not come on to the market until Japanese component companies expanded into Europe and America at the end of the 1970s, a development that heralds the next stage of the modern bicycle's biography.

RIGHT Campagnolo down-tube gear shifters from a 1978 Colnago bike. The Italian frame builder Ernesto Colnago engraved the levers with his company logo.

FAR RIGHT TOP Campagnolo's Super Record brakes are shown on the same 1978 Colnago bike as above. The bike was ridden to a number of victories by the Italian racer Gianbattista Baronchelli.

FAR RIGHT BOTTOM A 1995 Campagnolo Chorus rear mech on a race bike made by Dutch manufacturer Gazelle. Campagnolo was chasing Shimano for innovation at this time, and the Italian company had lost some of its design flair. Their components were less elegant during this time than they had previously been and soon would be again.

The dynasty >

Campagnolo celebrated 80 years in business in 2013. Over the decades, the company has had just two leaders: Tullio Campagnolo, the founder, and his son Valentino who runs the business today.

Campagnolo are the perfect example of the manner in which racing drove bike design.

While celebrating the company's 80th anniversary, Valentino said: 'Whenever we develop a product, that product is addressed to racers. Campagnolo has to continue to develop the winning culture that we have had in the past. That comes through a very close and tight connection with the racing field. It means not just making products for racing bikes, but also products that meet the racers' demands. The company needs to be always in contact with racers, its teams and the mechanics, that is the winning approach for Campagnolo. It is essential because, while inside the company we have a testing department where we invest a lot of money, what happens on the road is very different to what you can try to realize in the testing lab.'

BELOW Valentino Campagnolo, son of founder Tulio Campagnolo, poses at the headquarters of the company in Vicenza, Italy. The company began in 1933 in a Vicenza workshop.

Shifting East >

Europe's grip on the cycle industry was ended by a boom in bicycle buying during the early 1970s in America. European manufacturers didn't have the capacity to keep up while Japan was both ready and able to meet the increasing demand with innovative products. Japanese company Shimano has become the best-known bike component manufacture in the world today.

Suntour, a lesser-known Japanese company, also entered the cycling world on the back of the boom. At the start, Suntour was the more innovative of the two and, in 1964, developed the world's first slant-parallelogram rear mechanism. It was cheaper than anything the European manufacturers made and many thought it functionally superior. The design kept the chain approximately the same distance from the sprockets through each shift, reducing the potential for chain-slip. In 1969, Suntour introduced the first index gear system (see page 54), where one click of the shifter moved one gear. Suntour also brought out a free-hub on to which the sprockets were held by a locknut.

Index gearing was easier to use and much more dependable than friction systems and, as it required the use of levers and gears from the same supplier, the system helped imbed the concept of the group-set.

With indexed gears, the rear mech (and the front mech today) move a fixed distance for each click. Manufacturers soon realized that if they spaced their sprockets slightly differently to their rivals', customers would also be compelled to buy sprockets, chain and hubs from them.

Suntour did not make a group-set at the beginning of the American bike boom; instead, they teamed up with Sugino and Dia-Compe – who produced chain-sets and brakes, respectively. It would turn out to be a disastrous decision. The arrangement meant Suntour had little influence over the design of the components supplied by its partners. As the market developed and companies such as Shimano refined their offerings, Suntour was left behind.

Shimano >

Today, Shimano have 50 per cent of the global bicycle component market. As with Suntour, the American bike boom gave Shimano a global start when European manufacturers could not satisfy the increase in demand. Unlike Suntour, Shimano decided to expand and diversify its own manufacturing capacity rather than outsource, giving it a huge advantage as the market developed.

Shimano also marketed its products better than anyone, including the Europeans. The company started supplying Tour de France teams in the 1970s and by the 1980s they were even equipping loyal Italian teams, right on Campagnolo's doorstep.

Shimano carried on from Suntour with indexed gearing (see page 54). They went on to develop the integrated brake lever and gear shift system.

The next bicycle boom, that of mountain biking (see pages 136–7), which also started in America, gave Shimano its next big break. With everything already in place to develop products quickly, Shimano was immediately the biggest player in the off-road arena. Family owned Campagnolo, still reeling from Shimano's impact on the road market, made a half-hearted attempt to compete in the MTB scene with its Record OR (Off Road) group-set in 1989 but it was short lived. At the time of writing, the Italian company has no MTB equipment in its portfolio.

OPPOSITE LEFT Suntour's Gran-Prix slant-parallelogram gear, first produced in 1964, was inexpensive and, in terms of shifting speed and reliability, out-shone anything made in Europe at the time.

OPPOSITE RIGHT The Suntour Competition gear, first produced in 1965, was the first index gear system, for which one click of the shifter meant one shift of the chain.

TOP In 1973 Shimano made its Tour de France debut by supplying the brakes, gears and chain-set, plus other equipment from their Dura Ace range, to the Belgian Flandria-Carpenter pro team.

ABOVE It was also in 1973 that the Belgian rider Walter Godefroot of Flandria-Carpenter won the first Tour de France stage for Shimano. This picture shows the Flandria rider Wilfried David winning a later stage in the race.

Chapter 8
Going Off-Road >

A rider gets ready for the BMX Vert final at the Montjuic Pool in Barcelona, the Spanish city that hosted the X-Games in May 2013. Montjuic stands above Barcelona and has long been a home for cycling in the city. The hill is used for amateur road races and has hosted stage finishes of the Tour of Spain and the 1973 world road race championships.

Off-Road Adventure >

For many children there is nothing better than bombing around the local woods on a bike, rushing down rutted trails and mastering tricky off-road climbs and bumpy descents. And the fact is, ever since the bicycle was invented, adventurous grown-ups have been riding off-road too – perhaps in an attempt to relive the thrill of their early exploits on two wheels.

Outside major towns and cities, early roads were often crudely constructed of hard-packed dirt and stones, so the first cyclists had to deal with near off-road conditions as a matter of course. Even then, however, there were pioneering cyclists who wanted to explore further.

One of the earliest organized off-road adventures was undertaken by a section of the US cavalry known as the Buffalo Soldiers Bicycle Corps. Officially, they were the 25th Infantry Regiment, which was made up of a racially segregated group of African-American soldiers. In 1896, they were given the task of testing the viability of bicycles as an alternative to horses for personal transport.

A group of soldiers was ordered to test bicycles in a series of cross-country rides. The last of these trials was a 3,058-km (1,900-mile) journey over very challenging terrain, from their base at Fort Missoula in Montana to St Louis, Missouri.

The man behind the bicycle corps experiment, General Nelson A Miles, had seen cycling at the indoor six-day track race at Madison Square Garden in New York and, impressed by the speed of the cyclists, immediately started using cycle couriers in an army role.

Buffalo Soldier bike challenges >
For their test, the Buffalo Soldiers rode iron Spalding bikes. Their first exercise involved carrying 34.5kg (76lb) of supplies and kit on a 203-km (126-mile) trip to Montana's Lake McDonald and back. *En route*, despite encountering deep mud, fierce gradients and mechanical mishaps, they made it through in good shape.

Their next test was a return journey from Fort Missoula to Yellowstone Park – a distance of 1,223km (790 miles) – in which the Buffalo Soldiers rode through mud, across fields, through forests, on rocky trails and over mountains. They survived and were deemed ready for the final trial – the journey from Fort Missoula to St Louis. After passing this final test, the bicycle was duly adopted by the American army.

An account of their incredible adventures, entitled *Iron Riders, Story of the 1890s Fort Missoula Buffalo Soldiers Bicycle Corps*, was written by George Niels Sorenson and published in 2000.

RIGHT The Buffalo Soldiers Bicycle Corps, made up of a racially segregated group of African-American soldiers, established bicycles as an alternative to horses for troop transport over long distances and rough terrain.

Spirit of adventure >

As well as showing the bicycle's possibilities in combat, the Buffalo Soldiers' cycling expedition ignited a long-lasting passion for off-road adventure on two wheels. At first, this kind of cycling was called rough riding in America, and rough stuff riding in the UK and Europe. In other places around the world, such as Africa or Asia, off-road cycling was simply part of every day life and wasn't given any special designation.

Most early off-road riders were primarily adventurers, people who wanted to use bicycles to reach new places and explore just what the machine was capable of. There were very few organized events. Fearing that road improvements would turn young cyclists away from the joy of riding off-road, the Rough Stuff Fellowship was set up in the UK in 1955, to promote the activity.

This spawned the creation of other organizations throughout the world. In 1973 the Adventure Cycling Association (ACA) was inaugurated in America, with its headquarters in Missoula, Montana. As well as promoting off-road riding, the ACA is credited with creating a number of famous cycling trails.

In keeping with the free spirit of off-road cycling, no specific bikes were advocated; people simply made or adapted their own. Off-road-specific bikes did not become a recognized category until the sport of cyclo-cross developed (see pages 132–3), followed later by the invention of the mountain bike (see pages 134–5).

TOP/ABOVE/RIGHT An empty road disappearing into the horizon is as much a part of the cycling adventure as conquering a mountain pass, forging a forest trail or negotiating a twisting switchback descent.

Cross-Country for Bikes >

In parks, woods, muddy fields, even in quarries and on beaches, winter cycling takes place all over the world. Inspired by cross-county running races, an activity usually undertaken in the winter, cyclo-cross started as a way for road racers to keep fit through the off-season. Sport being sport, soon there were cyclo-cross championships. As this new category of cycle racing took root, racers began to specialize in the discipline of cyclo-cross.

Shortly after road racing took hold in Europe, some French riders developed a type of race called a steeplechase, in which they raced from town to town taking as direct a route as possible. Roads were used where possible, but riders also crossed fields – navigating by church steeples – to keep going in a straight line. They made ground in any way they could, running with their bikes on their shoulders where conditions did not allow riding.

Such cross-country races were formalized in the early 1900s. To allow organizers to attract crowds, the action switched to circuits. The first French national cyclo-cross championship was held in 1902. The trend spread through the rest of Europe, with the first Belgian title won by triple Tour de France winner, Philippe Thys, in 1910. Other countries quickly followed suit with their own cyclo-cross championships. As the sport grew in popularity, races were held in towns and parks, using natural obstacles – and where they weren't available, low walls and steps were incorporated to make the riders dismount.

In the 1950s, specialist cyclo-cross bikes began to appear. As road-race bikes of the era had good clearance between tyres and frame to stop mud clogging the wheels, some riders chose to stick with their standard machines. In fact, the 1960s cyclo-cross bikes were very similar to road-race bikes, the only significant difference being the use of cantilever brakes (see page 144). Later, cyclo-cross bikes were fitted with smaller

RIGHT Action from a 1942 cyclo-cross event that took place on the streets of Montmartre in the centre of Paris. The main obstacles were several sets of steps, like these leading towards the Sacré Coeur, that the competitors had to run up and down.

OPPOSITE LEFT A competitor in the same 1942 Montmartre cyclo-cross negotiates a drop-off outside the Moulin de la Galette.

chain-rings, to provide lower gears and a wider range of sprockets on the freewheels. Aggressively knobbly tyres, developed by the 1950s, completed the ensemble.

Cyclo-cross tyres became more sophisticated, with specialized tread patterns emerging to maximize grip on different surfaces (see pages 152–3). Muddy conditions demanded a widely spaced, studded tread, to grip and clear mud easily. In contrast, sand tyres were fatter, with a smaller tread, and rocky or hard conditions required a finer tread pattern again. Tyre lore, a mix of fact and beliefs, is still a constant topic of debate.

Belgian brothers >

Although they were both world cyclo-cross champions, Erik De Vlaeminck was just a little better than his younger brother Roger. He won the professional title seven times between 1966 and 1973. While Erik is regarded by many as the greatest cyclo-cross rider ever, Roger had the edge in road racing. In the 1970s, he won many of the world's biggest road races, including Paris-Roubaix four times. The two were so exceptional at off-road riding that they were often able to keep riding while others had to dismount and run. Long after cyclo-cross bikes had become the norm, Roger De Vlaeminck would often use a lighter road machine, shod with cyclo-cross tyres, in order to go faster.

BELOW Brothers Roger (left) and Erik De Vlaeminck, world cyclo-cross champions, in 1969.

Repack >

Mountain bikes saved the cycle industry when it was going through a recession. They helped draw thousands of young people towards cycling at a time when they were surrounded by other emerging attractions. The child-like thrill of swooping through the woods, hammering along trails, skidding through corners and getting air off bumps and berms (banks or ledges of soil) appealed just as much to grown-ups. Debate over who actually invented the mountain bike still rages on.

Off-road riding, although not recognized as a distinct activity at the time, goes back to the beginnings of cycling (see pages 130–1). The birth of mountain biking as a distinct category can be traced back to two specific places in the United States.

Two groups of cyclists – one in Cupertino, California, and the other in Crested Butte, Colorado – adapted cruiser bikes (see pages 100–1) to ride their local trails in the mid- to late 1970s.

To cope with the terrain, they fitted their machines with better brakes and fatter tyres but cruisers were heavy, so they focused on downhill riding. Inevitably, their exploits became competitive.

One of the earliest downhill events, the Repack, was first held in 1976. The name was a reference to the drum brakes that were used on early bikes. These were being used so heavily on the long downhill trails that friction inside the brake drum burned off all the grease, requiring them to be repacked with grease after almost every run. Repack races were held on Pine Mountain near Fairfax, California, and the adapted bikes became known as klunkers.

Gary Fisher and Charlie Kelly >

Gary Fisher, a road and track racer in the 1960s, was suspended from competing for breaching a rule about competitor's hair length. The regulation was withdrawn in 1972 and Fisher, still with long hair, took up cycling again. With a Schwinn Excelsior

LEFT An early 1980s mountain bike being carried across snow at 3,621m (10,700ft) on the Schofield Pass, near Crested Butte in Colorado, by US Governor Richard Lamm.

OPPOSITE TOP LEFT The Stumpjumper, first produced by Specialized in 1981. Its clean lines helped to make the bike an immediate success.

OPPOSITE TOP RIGHT Gary Fisher is one of the American pioneers of mountain biking. He took part in the first Repack races, then founded his own 'Gary Fisher' mountain bike brand. It was taken over by Trek, who now produce a range of mountain bikes called The Gary Fisher Collection.

OPPOSITE Early mountain bikes tended to be over-engineered and nothing like the sleek thoroughbreds of today. The steel frame of the Breezer 1, the first-ever mountain bike, was reinforced with extra tubes.

bike that he adapted himself, he took part in the Repack races and was very successful. He still holds the course record of 4 minutes and 22 seconds. Fisher's roommate, Charlie Kelly, was the Repack organizer. He coined the phrase 'mountain bike' when he got frame-builder Joe Breeze to start making bikes to his specification.

In 1979, Kelly, Fisher and another rider – Tom Ritchey (see page 137) – set up MountainBikes. It was the first company from which people could buy this new kind of bike. Kelly later sold his share of the business to Fisher, who renamed the company Fisher Mountain Bikes in 1982. Meanwhile, Kelly set up the first ever mountain bike magazine, *The Fat Tire*, in 1980.

Specialized >

The first full-production mountain bike – that a customer could go into a shop, buy and ride away on – was produced by Specialized: the Stumpjumper. Specialized still produce the model today, although it is a much more refined machine than the original.

Specialized's founder, Mike Sinyard, had the first Stumpjumpers made in Japan. The bike was formed around a welded steel frame, steered with BMX handlebars and driven via 15-speed Suntour gears, which were married with a touring bike chain-set and brakes. The finished package weighed a hefty 13.64kg (30lb).

The Mountain Bike Boom >

Between them, Gary Fisher, Tom Ritchey and Specialized set off a boom in 1981. The first 350 Stumpjumpers from Specialized sold in a week. It was marketed as a 'bike for all reasons'. Other American companies started producing mountain bikes, and the racing focus widened from just down hills to encompass all terrains.

As cross-country mountain biking grew, both as a sport and as a personal challenge, other established bicycle manufacturers included mountain bikes in their range. At the same time, new companies started up specifically to take advantage of the emerging trend.

The basic design was a welded steel frame – commonly made from Taiwanese Tange tubing – with 26-in aluminium wheels and fat knobbly tyres of around 5cm (2in) in width. Motorbike-style brake levers were mounted on flat handlebars, together with thumb shifters, which operated derailleur gears working with a triple chain-set and five, six and then seven sprockets on the hub. To cope with often steep hills, the sprockets were widely spaced, the biggest having at least 28 teeth. The chain-sets ring sizes varied, with 46, 34 and 24 teeth being popular. A bottom gear, usually 34 x 28 or smaller, was referred to as a 'granny gear' because it was so easy to pedal. Early brakes were either cyclo-cross cantilever style or U-brakes.

Mountain bikes could negotiate lots of different types of terrain, but they were heavy. As racing developed in North America and then in Europe, the search for lighter materials resulted in the development of aluminium frames. With an eye on race performance as much as rider comfort or safety, manufacturers developed suspension systems and better brakes.

Manufacturers such as Kona and Cannondale grew on the back of the American mountain bike boom to become cycling

household names. The early pioneers are also still going strong: Joe Breeze made some of the lightest mountain bikes during the late 1980s and early 1990s, and the Gary Fisher brand is still one of the best known in the world. But one name has become truly synonymous with the history of the mountain bike: Tom Ritchey.

Tom Ritchey >

Originally a racer himself, Ritchey built his first bike frame in his parents' garage in 1972. To save weight, he experimented – successfully – with building custom steel frames with thin-walled, large diameter tubes.

After working with Gary Fisher and Charlie Kelly (see pages 134–5), Ritchey began building bikes for his own company, Ritchey Design. In 1983 he quickly expanded his interests into bicycle component manufacture. From 1983 and through the early 1990s, he pioneered the development of several game-changing components for mountain bikes, including clipless pedals, threadless headsets and a range of tyres.

Above all, Ritchey will be known for the fine tolerances of his work, which allowed him to use the lightest materials to create sophisticated frames and equipment. Ritchey's best work was branded WCS (World Champion Series).

BELOW A Cannondale Super V from 1994. This full-suspension bike used a rear shock absorber and Cannondale's Headshok suspension forks at the front.

OPPOSITE The Kona Cinder Cone was made of steel, with a balanced geometry that performed well off-road. The bike was long and low, and one of the first designs to look like a modern mountain bike.

BELOW RIGHT Mountain bike pioneer and master bike designer and builder, Tom Ritchey.

Mountain bikes in Europe >

They say that when America sneezes, Britain catches a cold. That certainly occurred with mountain bikes and it wasn't just Britain – the whole of Europe caught the bug. Young Europeans associated mountain bikes with the cool side of America, with surfer dudes and mountain adventurers riding Rocky Mountain trails. However, European trails were not at all like the Rockies. They were muddy. The American styling of mountain bikes needed a European tweak.

By the late 1980s, when mountain biking caught on in Europe, American mountain bikes were already being mass-produced in Taiwan. The first mountain bike to sell well in the UK was the Specialized Stumpjumper and it was soon followed by bikes from

British mountain bike manufacturers Muddy Fox, originally a BMX maker, and Saracen, set up in 1987. Both rushed to meet the fast-growing demand. Major manufacturers in the rest of Europe soon followed and launched mountain bikes with tungsten-insert -gas (TIG)-welded frames that had been manufactured in Taiwan. TIG welding lends itself to mass production because it is a much quicker process than brazing, which was the traditional way of joining steel tubes in bike manufacture.

During the late 1980s and early 1990s, a number of hand-built mountain bikes were made specifically for the British market by artisan frame-builders such as Overburys, Pace, Dave Yates and Chas Roberts. Two stood out – those by Overburys and Pace – as they bucked the trend of copying American bikes. These companies understood that the UK did not have the same

BELOW The 1982 Cleland Aventura, designed and built for European off-road conditions, was fitted with hub brakes and mudguards.

terrain as America and so their bikes were designed with greater clearance between tyres and frame and forks to cope with muddy British trails rather than dry rocky American ones. Another difference was that British artisan frame-builders tended to braze their frames and, because brazing uses lower temperatures than TIG welding, this allowed them to use thinner – and therefore lighter – steel tubes.

The man who really invented the mountain bike >

The old adage 'there's nothing new under the sun' is certainly true in mountain biking. Elements of every improvement, innovation and breakthrough in cycling can be found somewhere in previous pages of the cycling history books.

Geoff Apps was a motorbike trials rider. He competed in observed trails in the 1960s. These were outdoor tests of skill and precision, requiring the rider to negotiate a carefully laid out course over natural obstacles. Apps was not happy about the impact and damage these events, with their heavy, noisy, bikes, did to the countryside. In 1965, well before the 'repackers' (see page 134) he started adapting bicycles to ride off-road. By 1979, using Finnish snow tyres to cope with the British conditions, Apps began manufacturing off-road bikes.

Apps experimented with wheels of a greater diameter than 26-in – the size used almost exclusively on early mountain bikes – so was arguably the torchbearer for the 29-er mountain bikes we see today (see pages 110–11). The bikes were sold by Apps under the Cleland brand between 1982 and 1984.

In 1987, Apps co-founded mountain bike magazine, *Making Tracks* and now lives in Scotland, where he is still developing his own 'different' brand of off-road bikes.

ABOVE The D.O.G.S B.O.L.X, a 1992 creation in steel, from the London frame-builder Chas Roberts. A very modern-looking bike with an interesting name.

RIGHT Geoff Apps enjoying the British countryside with one of the mountain bikes he developed, the Aventura TT.

Soaking up the Bumps >

Early mountain bikes had rigid frames and forks (see pages 134–5). Their fat tyres absorbed some of the shocks and knocks from the rough terrain that the bikes were ridden on but tyres can only do so much to remove the battering a rough trail can inflict. Off-road bikes needed more – they needed suspension.

Front suspension >

Even a rigid bike has some suspension – the frame, components, wheels and, of course, tyres, all absorb shocks. Experiments and inventions designed to smooth the riding experience are almost as old as the bike itself but it was the advent of mountain biking that truly focused attention on the need for suspension.

As the front of the bike hit the bumps first and steered the bike, designers started by looking at this first. As well as being uncomfortable for the rider, rough trails could also dramatically affect steering and traction. The obvious solution to the problem was telescopic forks.

Initially, metal or elastomer springs were used in forks, then air was introduced as a lighter alternative. Metal, elastomer and

RIGHT A Cannondale mountain bike equipped with the company's patented Headshok suspension system, with a 'Fatty' fork. The working parts of Headshok are covered by a black rubber boot, which can be seen between the fork crown and the head tube.

ABOVE (Clockwise, from left) An early Rock Shox suspension fork; Cannondale's Lefty Max Carbon 140 fork; the workings of the Headshok on a Fatty fork removed from the bike.

air springs have different compression rates. Coil metal springs have a near linear rate of compression, while compressing gas can be made to act progressively.

At the beginning of the mountain bike boom (see pages 136–7), designers did not limit themselves to looking at forks. Handlebars were sprung with articulated stems: the Girvin Flexstem pivoted on an elastomer spring, reducing the vibration that passed through the bike and into the riders' arms. Early telescopic seat pins also provided some extra shock absorbency at the rear of the bike.

The innovative American bike manufacturer, Cannondale, has always looked at bike suspension differently. Its Headshok system, introduced in 1992, was a rubber-covered spring between the bike's forks and head tube. In 2000, after experimenting with the design for two years, Cannondale launched their Lefty fork; a radical single-legged fork with internal spring and damping.

Rear suspension >

The first bikes with front and rear suspension appeared in the early 1990s but tended to be heavy and provided an unnerving, bouncy ride, as they reacted to the rider's pedalling action. This effect – called pedal bob or kick-back – absorbed some of the power the rider put into the pedals. Early rear suspension also tended to compress under braking, severely affecting handling.

The answer for the masses came in 1992 when Gary Fisher launched the RS-1. It was the first commercially successful bike utilizing the four-bar linkage suspension system, designed by former motorcycle champion, Mert Lawwill, in 1991. Another, cruder, early rear suspension system, first seen on bikes in the early 1990s, was called the 'soft tail'. This used the flex located in the rear triangle of the bike as a spring and controlled it by putting a shock absorber – or sometimes just an elastomer – between the site of flex and the rider. The KHS Team Soft Tail and Trek STP are examples of soft tail bikes.

RIGHT A carbon-fibre version of the Cannondale mountain bike shown opposite. The Fatty fork is aluminium on this model. It is possible to 'lock out' the Fatty suspension fork, changing it to a rigid fork, by using a lever. This can be seen on top of the fork steerer above the bike's stem.

Evolution of Bicycle Suspension >

Modern bike suspension is a sophisticated business, and manages to provide good shock absorbency while also improving traction and controllability. There are currently full-suspension mountain bikes that weigh less than the best 1980s road-race bikes.

With a much wider range of potential applications, it is unsurprising that front suspension reached design maturity quicker than rear.

The Girvin Flextem (see page 141), with its limited amount of shock absorption and 'soft' feel, fell out of favour early in the story. Similarly, experiments with heavy, complex linkage systems in forks also fell by the wayside. Telescopic damping seemed to offer the best solution and quickly became the preferred development route. Light, air-sprung forks are now standard for cross-country mountain biking, while long-travel metal springs are favoured by downhill riders, who prefer the linear spring rate and a more robust fork.

Travel is an important term in suspension and refers to the amount a fork or shock absorber can compress when the bike hits an object. Cross-country mountain bikes have short travel suspension compressing 80mm-120mm whereas downhill bikes, which have to deal with higher speeds and bigger knocks, have long travel suspension. These can compress as much as 260mm. Trail bikes sit between the two extremes and typically utilize forks with suspension travel in the 120mm-140mm range. They are the most popular MTB.

Fork refinement might have happened first but rear suspension, using a myriad of different linkage systems, soon caught up. The main objective of all suspension systems is to absorb shock. The best systems react to the trail but not to the rider pushing on the pedals or pulling on the handlebars. To do this, the rear triangle – the part being suspended – is separated from pedal input at the bottom bracket by multiple linkages, or by placing a single pivot above the bike's bottom bracket.

ABOVE Suspension forks on an early safety bicycle. The fork end pivots around a bolt, while suspension is provided by a spring that can be seen above the footrests.

RIGHT A full-suspension bike that is suitable for cross-country racing and general trail riding.

BELOW Rear suspension systems are defined by the location of their pivot points on the bike. This bike has a single pivot point mounted to the down tube.

RIGHT One of the latest suspension forks produced by Rock Shox that incorporates the fork legs with the fork crown as one piece of carbon fibre, thus reducing weight. The stanchions, to which the wheel is attached, slide up and down inside the fork leg. The exploded diagram shows the component of the suspension and its damping system, which are also inside the fork leg.

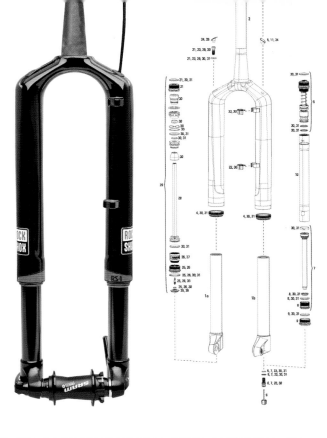

LEFT A downhill race bike from Specialized. Rock Shox BoXXer forks provide huge suspension travel to absorb big bumps at the front of the bike, while the spring rear suspension can be fine-tuned (inset) to suit any rider and course.

Evolution of Mountain Bike Brakes >

Everything about the off-road environment magnifies the demands on equipment. Brakes must have strong, progressive and precise off-road stopping power and they must work well when brake surfaces are wet, dusty or covered with mud.

Because there was really no alternative, the first off-road machines were forced to use standard road brakes. Later on, more powerful cyclo-cross cantilever brakes were used. Then, as the off-road market grew, MTB brakes developed from cyclo-cross cantilevers into a whole new line of off-road brakes. As time went on, disc brakes that had been developed for mountain bikes and off-road cycling were adapted and fitted to some road bikes.

Once mountain bikes went into mass production, they were fitted with cantilever brakes. These are centre-pull brakes with two spring-loaded sides, independently mounted on to metal bosses, welded to the seat stays and fork (see page 122). Originally designed for tandems (see pages 62–3) and trikes (see pages 60–1), cantilevers are very powerful. An additional factor that made them popular for use on cyclo-cross bikes (see pages 132–3) was their open design. This allowed mud, picked up by the tyres, to pass freely between the callipers.

This good clearance feature adapted well to the fat tyres of mountain bikes, allowing easy removal and fitting of wheels.

As mountain bike design developed, it exposed a major drawback with cantilever brakes. The power of a cantilever brake is determined by the length of each brake arm, but long brake arms on cantilevers stick out too much and they can snag things, like mud, leaves or even moving suspension parts. V-brakes were developed to get around this problem. With these linear side-pull versions of cantilever brakes, the cable pulls from the side and the two brake arms are in a vertical plane, so they don't protrude. That keeps them within the shape of the bike frame, while the arms are long enough to give plenty of brake power.

V-brakes were slim, powerful and easy to adjust. As their longer arms required a lot of brake lever pull (*pull* being the term to describe the clearance between levers and handlebars) they only worked with mountain bike brake levers. Road bike brake levers did not have as much clearance, so not as much pull.

V-brakes worked so well on mountain bikes that it would be a decade before the now common cable disc brakes were adopted. Although companies such as Hope Technology in the UK were developing cable disc brakes as early as 1992, they were a heavier system that did not offer much of a performance gain.

RIGHT A modern disc brake calliper and rotor, the Avid CODE™. The brake pads are inside the minimalist calliper body and are pushed by hydraulics on to the rotor disc to brake.

First appearing in the early 1990s, hydraulic rim brake systems were both easier to use and more powerful than their cable-operated brethren. However, like the cable disc brake, they did not offer enough of an advantage to justify their extra weight and complexity, so never gained wide-scale acceptance. In 1997, the American company Hayes discovered the magic formula when they combined these two technologies to create the hydraulically operated Hayes Mag disc brake. It was a winning combination and the cycling world took a leap forward in stopping performance.

Disc brakes >

In wet and muddy conditions – frequently encountered in mountain biking – rim brake performance dropped dramatically. Even in warm, dry conditions, rim brakes were susceptible to performance change. Long descents, with constant braking, caused heat build-up and sometimes brake fade. In contrast, harder brake pads and steel construction made disc brakes less susceptible to all of these things. In addition, the positioning of disc systems (away from the road surface) made them almost impervious to wet and muddy conditions. At last, mountain bikes had a system that could be relied on in the harsh, mountain environment.

Disc brakes on road bikes >

Although hydraulic disc brakes were developed for mountain bikes in the 1990s and have gained wide acceptance, the Union Cyclist International (UCI) has not permitted their use in road events. Because public buying habits are strongly influenced by what the top racers use, manufacturers have been reluctant to build disc-equipped road bikes. Since 2010, bike makers have identified a demand among buyers for hydraulic disc brakes, so some manufacturers supply them on some models across their ranges, regardless of whether pros can use them or not.

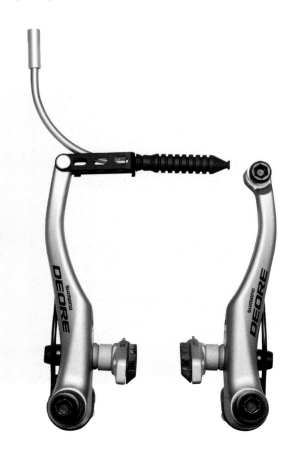

ABOVE The Avid CODE™ hydraulic disc brake lever is a great example with its calliper (opposite) of how light, simple, yet still very powerful bicycle disc brakes have become.

RIGHT A Shimano V-brake calliper. V-brakes are a development of the cantilever brake specifically designed for flat handlebar mountain bikes. They are easier to adjust and their cables are simpler to fit than cantilever brake cables, but they have similar stopping power.

Downhill Racing >

As they race down narrow tracks, weave between trees, over boulders and vertical drop-offs, riders performing on a top-level downhill mountain bike run are an incredible spectacle. Downhill courses can be fun and exhilarating but are definitely not an activity for the faint-hearted.

At the first mountain bike world championships in 1990 there was both a cross-country and a downhill race. In those early days, many competitors competed in both events and on the same bike. However, as the sport progressed, bike designers responded to the more specific demands of each discipline. The geometry and suspension system of modern downhill mountain bikes became so specialized that it is only really suitable for this one activity. Between runs, rather than ride their bikes back to the summit, riders often use ski lifts or pulls.

The big bumps and jumps of downhill courses require heavy-duty, long-travel suspension front and rear. Bike parts are designed for strength rather than lightness, and the bikes are long and low to increase their stability.

Even when it is almost fully compressed by cornering forces, a mountain bike has to track steadily, absorbing both big and small bumps quickly, with little rebound (see pages 150–1).

Mountain bike centres >
Mountain biking trail centres have sprung up all over the world and, like ski centres, many have different grades of run that are designed to suit different abilities. Although some routes are entirely natural, this is rare. The majority of routes incorporate man-made features, such as built-up berms in corners, man-made drop-offs, jumps and bomb holes. These spice up the run and also help to control erosion.

Downhill bike racing has spawned several different offshoots, among them dual slalom, in which two riders go down the course at the same time, and four-cross, in which four riders compete at once. The courses for these events have a mix of natural and man-made obstacles on a steep descent. Races last typically less than a minute and are designed to be engrossing for spectators.

RIGHT A downhill racer leaving the start house on the Val d'Isère run. Downhill racing is the summer sport at many ski resorts. Racers wear special clothing with padding and full-face helmets.

OPPOSITE Natural hazards like trees and rocks feature on many downhill courses, like that being navigated here by a competitor in the American downhill title race in 2010. Riders are not allowed to deviate from the course, which is marked out here with blue tape.

Anne-Caroline Chausson >

There have been lots of great downhill racers, but none with such a spread of talent as the French downhill multi-world and Olympic BMX champion Anne-Caroline Chausson. As well as her BMX achievements between 1993 and 2005, she won three junior and nine senior world mountain bike downhill titles. Using both her BMX and downhill skills, she was twice dual slalom world champion and twice four-cross world champion.

Sadly, BMX is the only one of these disciplines currently on the Olympic cycling programme. In 2007 Chausson switched back to her first sport, BMX, and took the inaugural women's Olympic BMX gold medal in 2008.

Modern Cross-Country Bikes >

International cross-country courses have changed over the years, mainly in response to the huge cost of televising the discipline at the 2008 Beijing Olympics. To begin with, courses were long and largely natural. Now, they are often more compact and frequently incorporate man-made obstacles, such as log jumps, drop-offs and rock gardens. This change in challenges has affected cross-country bike design. As well as front suspension bikes (commonly known as hardtails), riders often opt to use lightweight, full-suspension cross-country bikes, to cope with the sometimes severe demands of modern, international-standard courses.

BELOW A modern cross-country race mountain bike (CBoardman Bikes). It has a carbon-fibre frame, carbon-fibre wheel rims and knobbly tyres with a tread pattern that gives good grip but allows mud to clear easily [1], an ultra-light saddle [2], forks with enough travel to cope with all cross-country race conditions [3] and disc brakes [4].

BELOW The Specialized S-Works – a modern full-suspension cross-country mountain bike – being put through its paces.

BELOW Another modern full-suspension cross-country mountain bike (CBoardman Bikes) with disc brakes and a wide range in sprocket size to cope with different gradients and conditions [1]. The shock absorber controls how much the rear triangle of the bike moves in response to the terrain being covered [2]. It also controls the speed of that movement. The shock absorber can be adjusted or even locked out completely while riding. Suspension forks absorb shock at the front of the bike and can be tuned specifically to work optimally with the rear shock [3].

Modern Downhill Bikes >

Downhill mountain bikes have evolved into highly specialized machines, able to cope with the extremely specific demands of the activity. This is a good example of a pro-level bike with such a specific design that it is only really suitable for riding downhill (see also pages 146–7). Each part has been designed to withstand severe abuse. Its suspension can be custom-tuned to the needs of the rider or for tackling the demands of different types of courses.

3 The rear shock absorber has to cope with heavy knocks, so uses a metal spring. The spring function can be fine-tuned by devices on the shock absorber itself, including the smaller cylinder, seen here, that controls rider-induced movement in the shock.

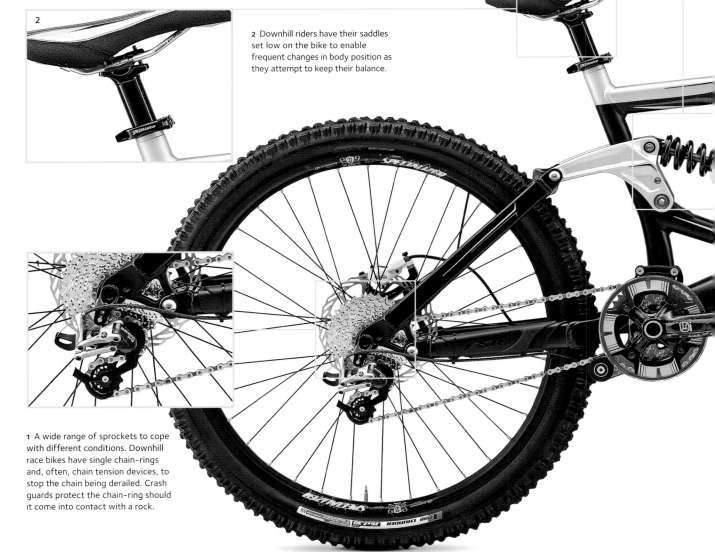

2 Downhill riders have their saddles set low on the bike to enable frequent changes in body position as they attempt to keep their balance.

1 A wide range of sprockets to cope with different conditions. Downhill race bikes have single chain-rings and, often, chain tension devices, to stop the chain being derailed. Crash guards protect the chain-ring should it come into contact with a rock.

4 There is no need for a downhill bike to be light, so substantial welded aluminium frames are used.

5 Robust forks with longer travel than those on a cross-country bicycle help the bike to cope with all downhill race conditions.

6 Powerful disc brakes provide plenty of easy-to-apply stopping power. The red knob at the bottom of the fork legs controls one of the adjustments that fine-tunes the forks.

Modern Cyclo-Cross Bikes >

Although road bikes were used in the first cyclo-cross races, their designs have now diverged too much for that to be possible (see pages 132–3). Cyclo-cross bikes are designed specifically for their task: racing fast over a cross-country course for one hour – the average length of a cyclo-cross race. The specific adaptations that make the modern cross-country bike a light nimble machine to ride over a wide range of terrain can be clearly seen here.

2 Cyclo-cross bikes need to be strong but light, so weight is saved everywhere, including with this cut-out at the rear of the saddle. This innovation also helps to stop mud from building up under the saddle.

1 As disc brakes are allowed in cyclo-cross competitions, many modern cyclo-cross bikes come equipped with them.

3 Internally routed control cables also help to reduce the build-up of mud on vital components.

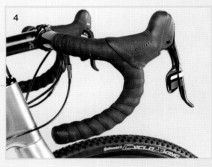

4 Cyclo-cross bikes traditionally have droped handlebars because the sport was invented as the winter alternative to road racing. This tradition is observed in the rules that apply to top-level cyclo-cross races, in which bikes must have dropped handlebars.

5 Cyclo-cross bikes have rigid forks. As cyclo-cross race courses aren't as extreme as some mountain-bike courses, there is little need for suspension.

BMX Race Bikes >

Bicycle motocross (BMX) racing is not just a great way to get kids cycling; it's also an Olympic discipline. Many Tour de France stars – particularly some of the better sprinters, such as 2002, 2004 and 2006 points jersey winner Robbie McEwen of Australia – started in BMX. So did many mountain bike and cyclo-cross racers, such as 2005 and 2013 world cyclo-cross champion Sven Nys, who was eight times the Belgian BMX champion. The environments might be different but the best BMX riders have similar physical characteristics to the sprinters competing in velodromes. BMX racing is all about explosive power, skill and, above all, nerve.

The demands of BMX racing >

Races are held on prepared tracks that twist and turn from a start gate through a series of banks, berms and jumps to the finish. Races typically last about 45 seconds. The start gate is on a steep downhill so riders get up to speed quickly. Competitions are made up of a series of knockout heats, called motos, and a final. A maximum of eight riders contest each round. Most BMX bikes have 20-in wheels but there is also a 24-in wheel cruiser class. Olympic BMX is restricted to 20-in wheels. Top-level BMX bikes, like the Intense Sonic below, have aluminium frames, carbon-fibre forks, a single gear freewheel and calliper brakes.

1 The small aluminium frame of a BMX race bike is light but very rigid for the instant power transfer required for quick acceleration. This bike has platform pedals but most BMX racers use clipless pedals. The saddle is set much lower than the handlebars because BMX racers rarely sit in the saddle.

2 Rear view of a BMX race bike, with the inset showing how the rear wheel is fixed in the frame, and the extra security that holds it there. You can also see the single V-brake used to stop the bike.

3 Front view of the same bike showing the handlebar shape with the strengthening bar. The inset shows the hefty four-clamp stem. BMX racers are like track sprinters: the best of them generate enormous power for a few seconds, which means that their bikes have to be exceptionally strong.

BMX Stunt Bikes >

Agile and robust, BMX bikes are ideal for tricks and stunts. Experts can ride up and down steps, slide along rails and pivot on their front wheel. Whereas BMX racing is an Olympic sport, BMX stunt riding has the X Games, an annual sports event organized by US sports broadcaster ESPN that focuses on extreme sports.

There are four disciplines in the X Games, and although discipline might not be the first word that springs to mind when thinking of young, free-spirited BMX stunt riders, it is exactly what is required to excel in any of the X Game events. The four disciplines are: Vert, Park, Street and Big Air.

Vert >

This discipline takes place in a half pipe, which riders use to gain the momentum to perform stunts while airborne. They will also hang on the lip of the half pipe to do grinds and stalls. The X Games half pipe is 8.2m (27ft) high.

Park >

In this discipline, riders perform tricks on ramps and jumps that are similar to those found in skate parks. The tricks are comparable to those in the Vert, but the emphasis here is on a flowing performance that transitions smoothly from one part of the park to another.

BELOW Action from a 2006 Freestyle Vert competition in Los Angeles. The rider is performing a trick where he is in front of his bike before re-mounting and landing (hopefully) safely.

RIGHT Another rider in the same competition leaves his bike mid-air. The half pipe can be clearly seen behind him.

Street >

Street riders use urban architecture found in public spaces to perform tricks. Street in the X Games uses a mix of rails, walls, steps and jumps to replicate the urban setting.

Big Air >

A Big Air course is like a ski jump with twists. Riders descend a giant ramp to gain speed, then chose one of three take-off ramps. Once in the air, they attempt a variety of tricks, including 30-degree spins, multiple spins or somersaults – sometimes only barely retaining contact with their bike. The winner is the rider who does the most complex jump.

Stunt BMX bikes >

Although the stunt bike bares a strong resemblance to the BMX race bike (see pages 154–5) there are several differences. Frames and forks have thick, steel walls and are built for strength rather than stiffness or low weight. For the same reason, the 20-in wheels tend to have more spokes and the tyres are fatter – 5cm (2in) for increased control. One major difference between BMX stunt bikes and all other bikes is the chain-set, which is mounted on the right-hand side. This is to allow riders to do slides along rails or the edge of walls using the left side of the bike.

BMX bikes used in Flatland – where tricks are done on flat paved areas – have metal pegs on the end of each wheel axel called stunt pegs. Riders stand on them to perform tricks. The flatland BMX tends to have a very short wheelbase, to better allow the rider to spin the bike on one wheel.

BELOW Simone Barraco of Italy competes in the 2012 X Games Street Finals, held in Los Angeles, California.

BELOW The Big Air Final in the 2013 X Games, also held in Los Angeles. American Kevin Robinson competes, upside down in mid-air. The numbers on the column behind give spectators some idea of the height that each competitor attains.

The Hybrid >

Part of the attraction of the mountain bike, and part of the reason why it has brought so many people into cycling, is its ease of use. The upright position is comfortable, as are the flat handlebars and easily accessible shifters. Like the mountain bike, hybrids tend to have a wide range of gears, allowing the rider to climb even the steepest of hills. The wheels and frame, however, are more closely related to the road bike and roll easily on tarmac.

Most hybrids do not come with knobbly tyres or suspension, as it is designed for small sorties on easy tracks and trails, not for full, off-road excursions. The hybrid is, as the name suggests, a compromise – a bike not fully attuned to any specific terrain. Instead, it allows the rider to do a little of everything.

Horses for courses >

Hybrids come in a lot of shapes and under a lot of names. Commuter bikes, for example, combine elements of road and mountain bikes (see pages 104–5), while cyclo-cross bikes (see pages 152–3) which work well on- and off-road, can also be considered a form of the road bike.

The original hybrids appeared in the late 1980s and early 1990s. French manufacturers Peugeot produced one of the first. It had mountain bike handlebars and gears married with 700c wheels, cyclo-cross tyres and a road frame.

Nowadays, some hybrids are biased towards road use; practical straight-handlebar bikes fitted with road tyres and aimed at commuters. Others, with heavy treads and 50mm suspension forks, lean more towards the mountain bike end

ABOVE Some hybrids have a lot of mountain bike in their design. This one would be suitable for anyone who wants a bike to perform okay on the road, but be up for some off-road fun too.

of the spectrum. While not intended for off-road use, this particular blend of features would at least allow the rider to take some detours on to slightly bumpy paths and tracks.

BELOW This mountain bike is a real all-rounder that would work equally well on- and off-road. Manufacturers also make hybrid bikes that lean more towards road riding. Such bikes wouldn't have suspension forks or tyres with grippy treads like this one.

Bikes in Service >

Since the turn of the 20th century, police officers, particularly those with responsibilities in large rural areas, patrolled on bicycles. A bike increased the range an officer could travel and provided an excellent vantage point. He or she could see over hedges and into gardens – no small advantage for a patrolling officer. During the 1960s, bikes went out of favour and police forces turned towards cars and vans for patrols.

Now, police bikes are back in fashion, this time for urban policing, in many countries. As well as being cost effective and environmentally friendly, bikes are able to thread through traffic and allow the user to take advantage of cycle-lane short cuts. Often the cycling officer can respond to an incident faster than a colleague in a car or on a motorbike.

A number of manufacturers produce police bikes. The Fuji Special, which is used by police departments all over the United States, is sometimes equipped with blue and red flashing lights, 27 gears and front suspension. It even has a siren. The International Police Mountain Bike Association (IPMA) is the organization that oversees bike training for American police and paramedics.

Medi-bike >

All over the world, paramedics and other medical first-response units are increasing their use of bikes for the same reasons as the police. A bicycle-mounted paramedic can often be on the scene quicker than an ambulance and its crew: this is a huge advantage when a fast response can mean the difference between life and death. In London, medical responders take advantage of the load-carrying capacity of the bicycle to get oxygen, defibrillators and other specialized life-saving equipment quickly to the scene of an incident.

ABOVE Two British police officers patrol the countryside on their 'roadster' bikes in the 1920s. These police bikes were made from steel, had upturned North Road handlebars and were equipped with a single- or three-speed gear and rod brakes. The bikes also had chain-cases to completely enclose the chain. Raleigh and BSA Cycles were two of the biggest manufacturers of police bikes in the UK, which are very similar to the Flying Pigeon bikes (see page 121) that are still made in China today.

ABOVE A Modern British police bike. Bicycles have always been used by the British police and they are perfect patrol vehicles for the modern urban environment. Not only do cycling officers have great range, but they can often respond quicker to an incident than those patrolling in cars. It's also easy for a cycling officer to engage with the public.

RIGHT Nick Mars, paramedic, working the streets of London in 2002. His bicycle unit was set up to treat people with sudden non-life-threatening illnesses or injuries.

A number of manufacturers produce police bikes. The Fuji Special, which is used by police departments all over the United States, is sometimes equipped with blue and red flashing lights, 27 gears and front suspension. It even has a siren.

RIGHT Fuji make a range of police bikes with different specification levels. Each one is designed to cope with heavy use and is easy to maintain.

Ranger Dave >
Bikes are now being used for patrol duties in national parks throughout the world. The National Park Ranger Service in America uses mountain bikes to patrol otherwise inaccessible areas. For an insight into the life of a cycling ranger, you can visit the blog of San Diego-based, Ranger Dave (http://sdrprangerdave.blogspot.co.uk/).

BELOW Rangers patrolling the parks of North America. Bikes are perfect for this work as they have little impact on the environmant and can adapt to almost any kind of terrain.

Chapter 9
The Revolution >

The disc wheel was one of the biggest steps towards improving bicycle aerodynamics. Disc wheels are also very stiff, so they don't soak up much pedalling power. There are modern disc wheels that even act like sails on a yacht in crosswinds and use the wind to help propel the rider along the road.

Aerodynamic Drag >

Air resistance is a topic that has already been touched on in this book (see pages 34–5). It is the dominant force preventing a cyclist from going faster: a cyclist riding briskly along a flat road encounters an increasingly disproportionate increase in air resistance for every little unit increase in speed.

In order to lower their bodies and reduce the size of the hole they had to make in the air, early racers developed dropped handlebars. Tight-fitting clothing was also introduced in a rudimentary attempt to smooth air flowing over the body. These were largely intuitive solutions. The question of what was causing the resistance was largely left untouched until the mid-1970s, when the sports authorities of former communist Eastern Europe – as well as a number of individuals in the West – began to seek aerodynamic improvements to give their athletes an advantage in competition.

It was known that sheltering behind an opponent made things easier, as did sharing the pace – taking turns at the front while others sheltered behind – so that riders could go faster.

Track racers and time triallists used close-fitting silk jerseys in an effort to smooth out airflow. As early as the 1950s, some track racers covered their crash hats with clear plastic covers to decrease resistance. Twenty years later, road racers of the 1970s adopted cotton caps to do the same. In the 1960s, the French five-time Tour de France winner Jacques Anquetil even visited a wind tunnel to help him achieve the most aerodynamic cycling position. It is probably no coincidence that he became the best time triallist of his era.

By the mid-1970s, bike designers and sports scientists began looking at the aerodynamics of cycling. One of the many things they explored was wheels. In any given moment as a bike travels forward, the bottom of the spoke, where it fits in the hub, is

effectively static while the top is travelling at twice the bike's speed. For example, if a bike travels at 48km/h (30mph) the top of the spokes and the rims are travelling at twice that speed – 96km/h (60mph). Taking into account that resistance doubles with every unit increase in speed, traditional wheels, with their square rims and round section spokes, are very energy inefficient.

Minimizing frontal area >

Most of the early efforts to lower air resistance were focused on minimizing frontal area by reducing the actual size or amount of the object (such as spokes) punching through the air. Matters really progressed when people started to examine in a scientific way what was actually happening to the air as it passed over a rider. They considered not only size but shape.

Before it hits an object, air is flowing roughly in a straight line. As it encounters an obstacle – in this case a cyclist –

it is pushed around the sides, interacting as it does so with the surfaces of the bike and rider. On its journey around the cyclist, the air encounters abrupt shape changes, in the form of equipment edges and clothing creases. These sudden directional alterations force the air to break away from the surface and 'swirl'. Because it is has 'detached' from the cyclist, this spinning or turbulent air travels much slower, producing a pressure difference. This creates a 'sucking' or 'dragging' force on the rider, hence the term drag. Fighting against this drag accounts for roughly 80 per cent of a racing cyclist's total energy expenditure.

Reducing the size of the object creating this drag is beneficial, as is ensuring the air stays 'attached' to the rider and equipment as long as possible. The longer it stays attached, the less drag on the cyclist. To achieve this, manufacturers, athletes and coaches started to make smoother shapes with fish-like contours.

OPPOSITE LEFT Dropped handlebars were the first step in improving bicycle aerodynamics. The great Dutch racer Jan Raas is shown here in 1981 holding the bottom of his handlebars, thereby lowering his body and reducing his frontal area. Cyclists call this riding position 'on the drops'.

OPPOSITE RIGHT Tom Simpson wearing a clear plastic helmet covering on his way to a silver medal for England in the 4000-metre (4374-yd) individual pursuit at the 1958 Commonwealth Games, Maindy Stadium, Cardiff.

RIGHT The relationship between minimizing frontal area and increasing speed is amply displayed again by Jan Raas, this time in 1979. Note the difference in the angle of his elbows in this picture and in the picture opposite. There, Raas is cruising early in a race, whereas here he bends his arms to lower his upper body still further as he seeks to ride faster during a decisive part of a race.

Aerodynamic Drag (continued) >

Reducing wheel drag >

The first direct attempt to reduce drag was the introduction of the flat or bladed spoke, which presented less frontal area than the traditional cylindrical spoke. Improvements in the shape of the spoke's cross-section – from flat to elliptical – came next. In the early 1980s, manufacturers started to make wheels that did away with spokes all together.

Who came up with the first disc wheel is a matter of conjecture, but two companies – Renn in Germany and Mavic in France – began production of disc wheels around 1980 by joining rim and hub together not with steel wires but with two aluminium discs. These first disc wheels were heavy and noisy, but once they were rolling they were significantly faster on the flat than any wheel with spokes. Once carbon fibre appeared on the scene as a construction material in the cycling world, weight was dramatically reduced. Disc wheels became the standard for time-triallists and track-pursuit riders. They remain so today.

ABOVE RIGHT A selection of early aerodynamic disc and spoked wheels made by the French company Mavic. The Mavic Challenger was one of the first disc wheels but, being made of metal, it was very heavy. The next generation carbon-fibre Mavic Comete was much lighter and also stiffer.

RIGHT Cole front tri-spoke and rear disc wheels. Disc wheels are affected by crosswinds, so a front disc wheel changes handling and can make the bike difficult to steer and control. Tri-spoke and flat-blade spoked wheels are affected much less by wind, so they are used in road time trials. For indoor velodromes, however, where there is no wind, most racers use front disc as well as rear disc wheels in time-trial and pursuit races, where good aerodynamics are crucial.

Who came up with the first disc wheel is a matter of conjecture, but two companies – Renn in Germany and Mavic in France – began production of disc wheels around 1980, by joining rim and hub together not with steel wires but with two aluminium discs.

LEFT A close-up of a flat-bladed spoke in a Mavic Ksyrium wheel. Flat-bladed spokes are such an aerodynamic improvement that they are used in many race wheels today.

BELOW Bernard Hinault, one of the best male road racers ever, was a five-time Tour de France winner, and had many other major race victories to his name. Seen here in the 1986 Tour de France in the polka-dots of the Tour's King of the Mountains competition leader, Hinault rides an early aerodynamic low-profile bike, while wearing a one-piece Lycra skinsuit and one of the first aerodynamic helmets.

Aero touches >

Other early aerodynamic improvements were much smaller than the disc wheel. Among these were brake levers with cables that ran under the tape wrapped around handlebars, rather than travelling to the caliber through mid-air. In a crude attempt to shelter the calliper, some racers fitted the front brake of their time-trial bike behind the forks.

The skinsuit was the brainchild of Swiss scientist, Toni Maier-Moussa. Stretched tight over the body, the smooth Lycra material removed the rucks and creases associated with the standard jersey and shorts combinations. It was a significant step forward, so it is ironic that it was not Maier-Moussa's original intention to design it.

During the mid-70s Maier-Moussa was working on the development of an aerodynamic bike (detailed on the next page) and while he had the use of a wind tunnel for that project, he decided to test textiles too. To create a baseline from which to measure, he had someone ride naked in the tunnel. He found that Lycra was even more aerodynamic than human skin. Maier-Moussa went on to found the clothing giant ASSOS who still produce some of the most advanced clothing for cyclists today.

The Low-Profile Bike >

Once designers started to consider the area of size, shape and surface finish as one holistic aerodynamic puzzle, they examined wholly different shapes for the bicycles.

The race bikes of the early 1970s were a similar shape to those of the 1900s: they had dropped handlebars and a double-triangle frame with a top-tube running parallel to the floor. Geometry had been tweaked, making them more responsive and they now had gears, better brakes and were lighter. Each of these changes represented small evolutionary steps, but not game-changing leaps. Those were about to start arriving thick and fast.

For a flat time trial, the rider holds only the bottom of the handlebars, meaning that the top section is largely redundant. At least that is what Toni Maier-Moussa, founder of ASSOS, thought. In 1976, the pioneering Swiss scientist was among the first to bring together an exotic new material – carbon fibre – and aerodynamic thinking to create a revolutionary result. His creation had bullhorn-style handlebars that were bolted directly on to the forks, thereby doing away with the need for both the top part of the traditional bar and the stem.

In addition, in an attempt to smooth airflow, the cross-section of some of the frame tubes on Maier-Moussa's machine was teardrop shape rather than the classic round.

Daniel Gisiger of Switzerland raced a road version of Maier-Moussa's bike to victory in the 1981 Grand Prix des Nations time trial (the unofficial world championship of the time). Maier-Moussa had shown that, with the bullhorn-style handlebars, there was no need to have a high head tube to hold traditional bars in the right position. As a result, head tubes shrank. As it was still not quite low enough to get the tucked position that riders were seeking, an extra drop was gained by reducing the front wheel size. The low-profile bike was born, and 26-in wheels quickly became the norm for races against the clock. Some went even smaller with 24-in front wheels.

With the use of small head tubes and miniature wheels, the frontal area of the bike was significantly reduced. That low profiles were faster than standard bikes was a fact brought home when, in January 1984, Francesco Moser beat Eddy Merckx's world hour record. The Italian set a new mark of 51.151km (31.7 miles), adding a massive 1.72km (0.45 miles) to the legendary Belgian's distance.

RIGHT Swiss rider Daniel Gisiger races to victory in the 1981 Grand Prix des Nations time trial on a road version of Maier Moussa's bike, but without the bull-horn handlebars.

When people saw the bike and technology that Moser used to best Merckx 'The Cannibal's' mark, this was thought to be the reason for his success. Moser's record-breaking bike was like nothing people had seen before.

Francesco Moser's hour >

With a list of victories longer than any other bike rider, Eddy Merckx is unquestionably the greatest male cyclist of all time. When in 1972 he covered a distance of 49.431km (30.7 miles) to break the world hour record, it was thought the mark would never be surpassed. The man who surprised everyone by doing just that, was 1977 world road champion Francesco Moser.

When people saw the bike and technology that Moser used to best 'The Cannibal's' mark (the nickname by which Merckx was known), this was thought to be the reason for his success. Moser's record-breaking bike was like nothing people had seen before. He and his team had set aside traditional thoughts of what a bike should look like and instead focused on the demands of the event in front of them. It was a period that captured imaginations, including mine; in fact, it was the genesis of my own attempt on the world hour record nine years later.

Custom-made by craftsmen at his own factory, Moser's bike was produced from chromed steel tubing, some of it in an oval shape to reduce aerodynamic drag. Other tubes were curved to bring the rear wheel under the rider. The bike had front and rear disc wheels and a solid plate chain-set. For his record attempt in Mexico City, Moser wore a full-length skinsuit and an early version of an aerodynamic helmet. His attention to detail was so all-encompassing that he had the inner part of the track surface sprayed with a low-friction coating. Cycling had gone scientific.

RIGHT Francesco Moser training for his 1984 world hour record attempt in Mexico City.

The Low-Profile Time-Trial Bike >

Francesco Moser was not the only rider to utilize low-profile bikes against the clock (see pages 168–9). Riders from countries such as East Germany and Russia had been using low-profile bikes in track events since the late 1970s. In the 1-km (0.6-mile) time trial, individual and team-pursuit disciplines, cyclists from the Eastern block countries won many titles astride such bikes. It was not long before low-profile bikes were being used in the biggest cycle race of all: the Tour de France.

BELOW Bernard Hinault at speed in a 1985 Tour de France time trial. Hinault was one of the first to embrace the aerodynamic improvements available to racers during the early to mid-1980s. Here Hinault is riding a low-profile, time-trial bike with a disc rear wheel and spoked front wheel. His front wheel is smaller than the rear to help reduce frontal area. Hinault was also one of the first to use clipless pedals, as seen here.

The Weird, the Wonderful and the Just Plain Ugly >

The introduction of aerodynamic knowledge, low-profile bikes and carbon fibre precipitated an experimental boom among manufacturers. These are just a few examples of some extreme designs that were dreamed up to cheat the wind.

The wheelbarrow bike >

I came across the Battaglin bike below, which looks like the James Dyson-designed wheelbarrow that uses a ball instead of a front wheel, in the track centre at the 1986 world championships. It had been brought for approval by the Union Cyclist International (UCI). Unsurprisingly, it never received the endorsement by the governing body and quickly disappeared. It might have been less than beautiful but the concept of shielding the rider's legs from the wind was a good one. Had it been allowed in competition it may well have proved an advantage. How it would have handled is another matter!

Fork design >

Traditionally, fork blades are positioned to follow closely the form of a bike's front wheel. The proximity of moving and static parts creates a 'dam'. Because air is unable to squeeze through the gap, a pressure wave builds here, causing significant aerodynamic drag.

Many different designs have sought to minimize this effect. Some have taken the fork blades even closer to the wheel, forcing the air to treat both components as a single unit. Others, such as the bikes that were utilized by the Great Britain team for track events at the London 2012 Olympics, went in the other direction and took the fork well away from the wheel, effectively 'unblocking' the air dam.

There has been a lot of speculation as to why the Great Britain team went down this radical route. However, as I led the team – the Secret Squirrel Club – that came up with the bike design, I am not allowed to comment on the reasons for the choice in shape. The full rationale will probably remain under wraps for some years to come!

Research has found that the shape of a disc wheel's sides is very important in managing airflow, particularly in crosswinds, where well-designed wheels can act as a sail and actually create some forward thrust.

RIGHT The wheelbarrow bike might have worked aerodynamically but it was an aesthetic disaster. Note the gear shifters at the end of the dropped handlebars. This idea was first used in the 1950s and 1960s and was brought back on this Battaglin bike to avoid interruption to the airflow. The placement of the levers removed the need for the rider to take their hands off the handlebars when changing gear with conventional down-tube shifters. Integrated brake lever and gear shifter units would soon solve this problem in a much more elegant way.

Gitane Delta >

The French bike manufacturer Gitane developed a number of aerodynamic bikes, which they referred to as their Delta series. Bernard Hinault and Laurent Fignon both rode early Delta timetrial bikes when they won the Tour de France in 1978 and 1985 respectively.

Through the early 1980s, in an attempt to reduce drag, Gitane continued experimenting with various fairings as well as wing-section handlebars. A fairing was even attached to the rear of the saddle in an effort to manage the turbulent air behind the rider (although due to its very stubby design, it is unlikely this particular intervention worked).

Regardless of whether the company's experimentation was effective, its activities changed the look of bikes so much that the UCI banned any fairing that was added only for aerodynamic reasons and was not a functional and integral part of the bike.

Funny bikes >

The bicycle designed by Chester Kyle for the 1984 American Olympic cycling team was so radical that it was referred to by many as the 'funny bike'. Kyle's frames were made from teardrop-shaped aluminium tubing. Adding to the futuristic look, the bike had both front and rear disc wheels. The front wheel – 24in in diameter to enable the team-pursuit riders to tuck closer into the slipstream of their teammates – was constructed from two stretched Kevlar skins bonded to an aluminium hub and a graphite rim. The rear wheel was made by stretching graphite skins over a honeycomb disc made of Nomex fabric, which was then bonded with aircraft adhesive to a graphite rim and aluminium hub. Each bike was custom-built for its rider at a reported cost of US$20,000.

At a time when the Olympics were amateur-only, the previously all-conquering communist countries boycotted the 1984 Los Angeles Olympics. As a result, the host nation had their most successful Games ever, taking gold in the individual pursuit and silver in the team pursuit. On the road – also using road-going versions of Kyle's machine – they secured a bronze in the 100-km (62-mile) team time trial.

LEFT The Delta's inverted handlebars steered the bike externally to the head tube and cut frontal area. The flattened down tube and seat tubes also smoothed the airflow. A 26-in front wheel and a 700c rear wheel are other aerodynamic upgrades.

BELOW 'Funny bikes' in action at the 1984 Los Angeles Olympics. The American men's team riding them won a silver medal in the team pursuit event.

The Evolution of Pedals >

The first pedals comprised a simple flat plate on a tube that revolved around an axle. Some of these had leather loops to hold the rider's feet in place. Early experiments found that fastening a riders foot to the pedal increased cycling efficiency. The practice became popular even before multiple gears and good brakes were developed.

From the 1940s onwards, the soles of a cyclist's shoes – usually made of hard leather – had aluminium plates nailed to them. These plates had deep grooves that slotted on to the rear of the pedal cage, preventing the riders foot from slipping fore and aft, increasing efficiency. In conjunction with the straps, the grooved shoes allowed riders to apply force for a full 70 per cent of the circular motion, rather than the 30 per cent downwards portion that was possible without these additions.

Quill pedals >
By the time the Tour de France started in 1903, the quill pedal had been invented. Quill pedals consisted of a metal cage attached to a frame with a hollow centre. The hollow centre fitted over an axle, which the pedal revolved around on bearings. The axle was screwed into the bike's cranks.

At first, just toe clips were attached to quill pedals but toe straps that could be tightened and released were added in the 1920s. Not long after, riders started using shoe plates to make their feet even more secure on the pedals, increasing pedalling efficiency further.

The use of accompanying straps highlighted a flaw in the design of the pedal and shoes of the day. Early cycling shoes had thin soles and the metal cages dug into them. As the power riders produced had to transfer through the flimsy footwear, many riders complained of foot pain when riding long distances. Tight straps created pressure points, adding further discomfort.

Thicker straps and larger, more robust shoe plates helped to alleviate this. Another problem with the arrangement was that the system required the rider to undo the straps when they wished to put their foot down. This was a problem in crashes or if a cyclist simply forgot.

Clipless pedals >
Although the clipless pedal had been invented in 1895 by Charles Hanson, it was three-quarters of a century later, in 1971, when Cino Cinelli designed the M71. Strap-free pedals had arrived. The Italian's design was neat and simple to use

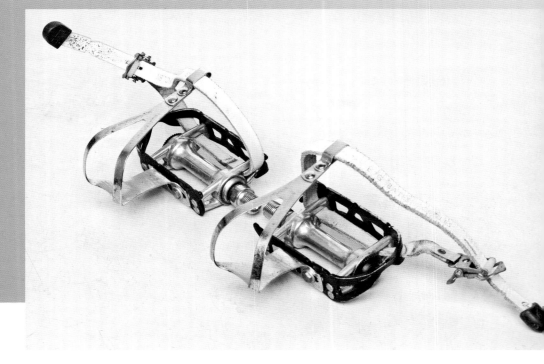

RIGHT Campagnolo Super Record pedals were the ultimate in quill pedals. Their light, black anodized aluminium front and back plates set them apart, and titanium axles made them very light. The metal toe clips and leather straps helped secure the rider's feet to the pedals. By using clips and straps, which allowed the rider to use 70 per cent of the pedal stroke – rather than 30 per cent without these features – pedalling became a lot more efficient.

but the true reason it succeeded was timing. Attaching plates capable of transferring rider power to the pedal without being torn free of the traditional leather sole was extremely difficult. Additionally, shoe uppers were not yet robust enough to replace tough straps and riders were reluctant to give up efficiency for convenience or comfort. It was the arrival of thermoplastics that made shoes stronger, stiffer and lighter. It also made clipless pedals a viable proposition.

Once combined with new shoe designs and tougher sole materials, the clipless pedal was quickly adopted, as it offered several advantages over the clip and strap.

The stiffer soles gave more efficient power transfer and helped spread the load over a wider area, while the lack of strap meant fewer pressure points. Once riders were familiar with the new system of engaging and disengaging their feet, it was a lot faster to remove a foot from a pedal in a hurry. As a bonus, in the event of a crash, clipless pedals usually released automatically.

Ski heritage >

The French ski-binding company Look was one of the pioneers of the clipless system. Its twist-release mechanism, first seen in 1984 and utilizing technology from its ski-binding business, was far superior to Cinelli's manual release. It quickly became adopted as the standard.

Just as for its ski systems, the Look pedals had sprung, 'hook-shaped' back plates. When the cyclist's foot pushed down, the rear lip on the patented shoe plate engaged and the foot was held solid. An outward twist of the heel was all that was required to release the foot from the pedal. The basic principal is still used by almost all pedal manufacturers.

There was only one drawback with early clipless pedals. Unlike clips and straps, which allowed the rider's foot to rock and twist slightly, Look's system held the rider's feet absolutely still. Many early adopters of the pedal started to experience knee pain as the required twisting motion was transferred up the leg. A new plate option quickly followed that allowed some lateral play, largely solving this problem. Today most pedals allow the user to custom-tune the degree of movement and retention strength.

BELOW An early Look pedal. The front of the cleats fixed to the soles of cycling shoes engage with the lip at the front of the pedal. The rider then pushes down with the heel, and the rear of the cleat clips into the black, spring-loaded retention mechanism on the pedal back.

BELOW Shimano XT clipless mountain bike pedals work in a similar way to the Look road pedals but the retention mechanism stands clear of the pedal body to allow mountain bike cleats to be recessed in the tread of the shoe. The rider can then run as well as ride over tough terrain.

BELOW Modern flat platform pedals are ideal for leisure use, for BMX riders who want to perform stunts and for downhill mountain bikers who frequently need to take their feet off the pedals. Platform pedals do not have retention devices.

Evolution of Cycling Shoes >

Like all footwear of the era, the first cycling shoes were lace-ups, made entirely from leather. When used for cycling, the traditional materials and construction caused several problems. The cage of the pedal bit into the soles of the shoes and the leather laces expanded when wet. In fact, the first cycling shoes caused riders so much discomfort that the footwear used in early super-long races was often exhibited to show just how tough this new sport was!

Road-race shoes >

The first step forward was the shoe plate. When fitted to the soles of cycling shoes – using traditional nails – the plates spread the pedalling load over a wider area, providing much-needed support. In the 1940s, some manufacturers, such as Italian company Detto Pietro, began incorporating metal plates into the soles of their products as standard. During the 1970s, to distribute the load further and reduce power loss, the Italian shoe manufacturer Duegi put wooden insole in its shoes.

The big breakthrough came later in the decade when Puma made the first modern cycling shoe with moulded plastic soles and an integral shoe plate. Plastic soles – and later carbon-fibre – were lighter, stiffer and supported cyclist's feet far better than leather. The new moulding techniques also allowed for the inclusion of threaded holes as standard, so shoe plates could be fitted without the need for nails.

The modern racing shoe is now highly sophisticated, with companies like Bont of Australia offering a fully custom-moulding service. Others, such as Lake, have a specially formulated sole that softens when heated, enabling riders to warm the shoes in their home oven, then mould them to the shape of their own foot.

The cycling footwear revolution did not neglect the upper part of the shoe. By the 1990s, most manufacturers had discarded laces in favour of velcro, ratchets and other tensioning methods.

Off-road shoes >

Early mountain bike racers used cyclo-cross shoes. These were effectively road shoes with the two spikes added just behind the

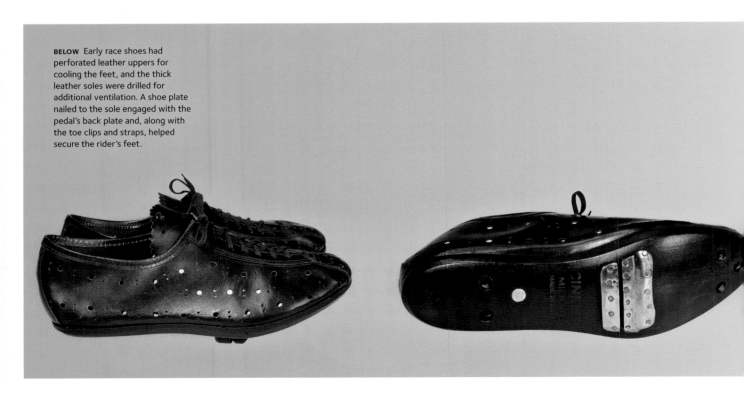

BELOW Early race shoes had perforated leather uppers for cooling the feet, and the thick leather soles were drilled for additional ventilation. A shoe plate nailed to the sole engaged with the pedal's back plate and, along with the toe clips and straps, helped secure the rider's feet.

slot that fitted over the pedal back plate. Off-road shoe design took a leap forward with the arrival of off-road clipless pedals (see pages 174–5). These used much smaller cleats (which were less likely to get clogged with mud than Look's design), recessed into heavily treaded soles. With the cleats tucked out of the way, the rider could walk or run easily on rough terrain.

BELOW With thick leather soles containing metal plates for the stiffness that was required, early leather race shoes were heavy. Wooden-soled shoes, as shown here, were lighter and stiffer.

BELOW Plastic-soled shoes were even lighter than wood and cheaper to produce. These Puma shoes from the 1980s came with an easy-to-adjust integrated shoe plate.

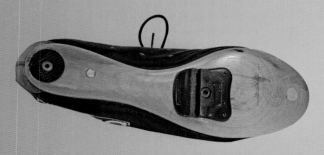

Everyday cycling shoes >

It is possible to ride a bike wearing almost any kind of shoe but the recessed cleat made walking so much easier that the system was quickly adopted by all recreational cyclists. Today, sophisticated cleat technology can be found in anything from custom carbon shoes weighing less than a sandwich to smart footwear that would not look out of place with a city suit.

Specialized Body Geometry Fit >

The American bicycle company Specialized has developed a comprehensive ergonomic bike-fit system, called Body Geometry Fit, that extends to customizing cycling shoes. The Body Geometry Fit mantra, 'Hand Foot Sit', encompasses the three main contact points of a rider with their bike – the handlebars, the pedals and the saddle. Processes have been developed that customize Specialized gloves, shoes and saddles – collectively called Body Geometry Fit products – to the specific anatomy of each rider.

People's feet don't just come in different sizes; they also come in different shapes. This is due to individual bone structure, muscular conditioning, injury and individual alignment. But the foot's contact with a bike, its pedals and cranks are all fairly standard. It makes sense then for cycling shoes to act as an interface between feet and pedals, taking into account individual foot variations, so every cyclist gets more or less the same contact with their pedals.

Specialized's Body Geometry Fit shoes achieve this by using different-shaped wedges and arch supports in different ways inside the customer's shoe so that the contact between the shoes and the pedals is ultimately the same for every rider.

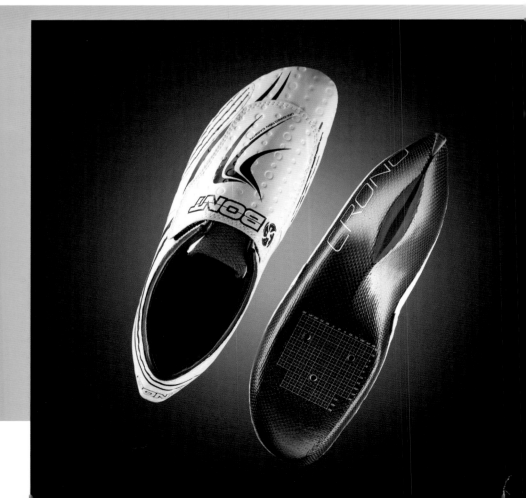

RIGHT Top-end race shoes, like these from Bont, have carbon-fibre soles which are both stiff and very light. These shoes were designed specifically for time-trial events, so they are aerodynamically shaped too. Shoes like these are built for cycling at speed. They are, however, difficult to walk in.

ABOVE Shoes suitable for mountain bike and cyclo-cross riding. Their metal cleats are recessed in a deep tread that grips the ground well. Off-road shoes with recessed cleats are also suitable for touring or for leisure cyclists who might want to dismount from their bikes and walk from time to time.

ABOVE The cleats on off-road shoes engage and disengage easily with off-road clipless pedals. These pedals have plenty of spaces to allow mud picked up on the cleats to squeeze through, keeping them clear to use. They also have retention mechanisms on both sides so that, regardless of which way up they are, the foot can engage.

RIGHT Speedplay is another brand of clipless pedal that allows the shoe to sit very close to the pedal axle. Most modern clipless pedals allow the rider's feet to pivot through each pedal revolution, which is claimed to be biomechanically more efficient than having feet held rigid. Speedplay pedals allow the greatest range of this movement.

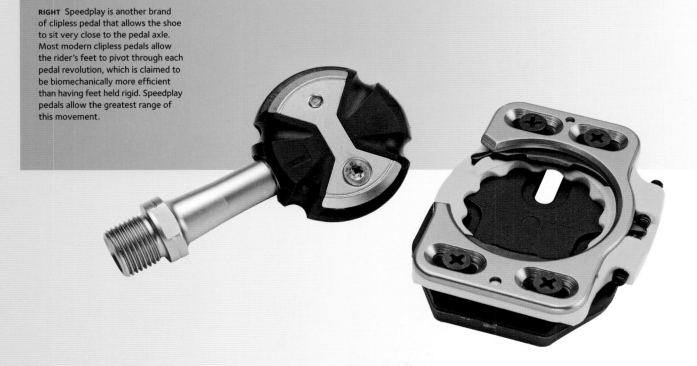

Quantifying Effort >

In the late 19th century and well into the 20th century, cycle training was based on historical practices, gut feeling and the crude observation of cause-and-effect. Coaches prescribed sessions in hours and distance to be completed, with stop watches and crude odometers providing the only empirical measurement.

Until the early 1980s, coaching was more of an art form than a science. But then two things happened simultaneously that turned coaches from qualitative to quantitative practitioners.

Measuring rides >

Early odometers counted wheel revolutions mechanically and provided a simple display. Speedometers were available, but they were heavy, clumsy to use and far from accurate. This changed when digital bike computers were introduced by American manufacturer Avocet, and Cateye from Japan, in the 1980s.

Other brands quickly followed and small, light, easy-to-use digital computers soon became a common bike accessory. They provided riders with information on how far they had ridden, as well as live and average speeds. Some models even displayed altitude and recorded the total height gained or lost on a ride.

In the 1970s an Italian sports scientist, Professor Conconi, noticed that there was a correlation between heart rate and chemical changes inside the human body in response to an increase in exercise load. Knowledge of this finding enabled cyclists to train with more accuracy. However, back then, monitoring the heart rate – and thereby quantifying an athlete's effort – was no

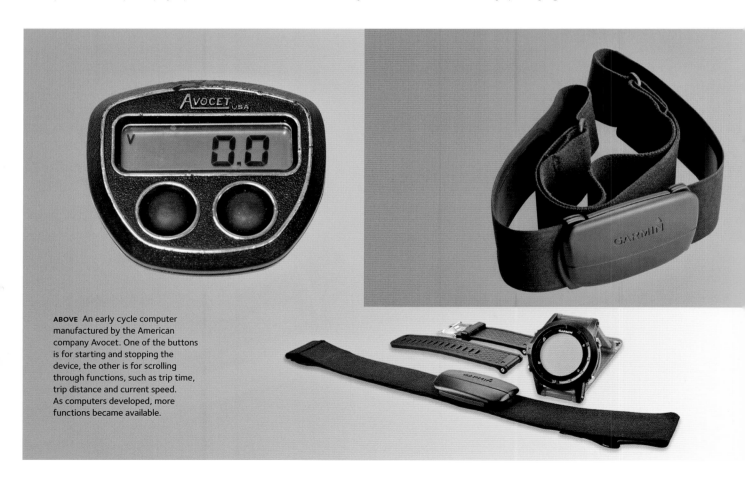

ABOVE An early cycle computer manufactured by the American company Avocet. One of the buttons is for starting and stopping the device, the other is for scrolling through functions, such as trip time, trip distance and current speed. As computers developed, more functions became available.

easy task. The only way to do so was to stop exercising in order to read the pulse, which was inconvenient and full of error.

Then, in 1977, Finish company Polar Electro introduced a portable heart-rate monitor (HRM) for general sale. An accurate way of constantly monitoring heart rate while exercising was available to sports people and coaches for the first time.

Heart-rate monitor >

This device consists of an elastic chest band housing electrodes that pick up the electrical signals from the wearer's heart. These are relayed wirelessly to a wrist or handlebar-mounted monitor. The heart rate is displayed digitally in beats per minute.

The first of these devices was bulky but, with a willing sports market to fund development, they quickly got smaller. As the size of the units shrank, their capabilities grew. By 1990, riders could correlate heart rate with speed and download training sessions to computers for analysis.

Professor Conconi developed a test that anyone with a heart-rate monitor could repeat. This test produces heart-rate zones for each individual. The zones are defined by a specific heart-rate range – 140 to 150 beats, for example – and training in each zone targets a different fitness capacity, from a sustained short-distance effort to the ability to perform pure endurance tasks.

For the first time coaches and athletes could formulate a training session using numerical data and predict exactly what effect it would have. The best among them knew how to do this before Conconi and HRMs, but only through intuition or (more likely) years of experience. For the rest it was guesswork and trial and error.

OPPOSITE/BELOW A heart-rate monitor showing details of the elastic chest strap that holds the pulse sensor and heart-rate transmitter in place. The watch-like monitor can be worn on a rider's wrist or mounted on the bike's handlebars.

BELOW The big breakthrough in quantifying effort when riding a bicycle came from the Italian sports scientist Professor Conconi, shown here with the great Italian racer Francesco Moser.

Aluminium Bikes >

Although a cheap material and extremely light when compared to steel, aluminium is not easy to work with, so it was many years before it became the widely accepted bike-frame material it is today. The principal challenge with aluminium was in creating strong and consistent welds. One of the early pioneers of alloy frames got around this problem by adopting an entirely different strategy.

Lodovico Falconi >

In the early 1970s, under the brand name ALAN, the Italian frame-builder Lodovico Falconi produced the very first mass-market lightweight aluminium bicycle frames. Rather than weld the aerospace-grade aluminium tubes together, Falconi chose to bond the frames using a high-grade epoxy glue and cast aluminium lugs.

To further capitalize on the new use for aluminium, ALAN frames had an anodized finish, which offered several advantages. Anodizing is an electrolytic process that increases the metal's resistance to corrosion and wear. It also creates a porous surface that accepts dyes well, so frames were dyed rather than painted. The resulting finish was lighter than paint and highly chip-resistant, thought their colour was prone to fading over time.

TIG welding >

Tungsten-inert-gas (TIG) welding is an arc-welding process in which a tungsten electrode creates the weld. The whole process is carried out in a bubble of inert gas. Advances in TIG welding during the 1970s made it even cheaper. Aluminium tubes used in bike construction are made from an aluminium alloy. Favourites for bikes were 6061 aluminium, which contains silicon and magnesium, and 7005, which is similar to 6061 but can be air-cooled to harden it (6061 cannot).

To start with, manufacturers simply rolled aluminium into traditional 25mm (1in) tubes but, as they became more familiar with the material, they realized that it could offer more. It was much easier to shape than steel, so lightweight, large-section aerodynamic tube profiles emerged. Companies such as Cannondale and Klein pushed further and started to experiment with increasing frame tube sections while decreasing wall thickness – a practice that yielded both lighter and stiffer frames.

RIGHT An early ALAN frame, made of aluminium with an anodized finish. This frame is the original aluminium colour. Even when manufacturers added a dye to the anodizing process, most left the joining lugs bare. This became a trademark of these glued aluminium frames.

LEFT Three examples of early TIG welded bikes from Cannondale. Above is a 2005 Cannondale Slice time trial/triathon bike. This aluminium bike is made with good aerodynamics in mind; the frame has flattened aluminium tubes to lower frontal area and smooth out airflow over them. In the middle is the 1996 Cannondale Super V. This aluminium bike was designed for downhill racing. It is equipped with front and rear suspension. Below is an early aluminium full-suspension cross-country mountain bike, the 1997 Cannondale Raven.

Carbon-Fibre Bikes >

Carbon fibre, as it is used in the cycle industry, is short for carbon-fibre-reinforced polymer. It is an extremely strong material in which a polymer matrix is reinforced with carbon threads. A polymer is a macromolecule that provides a framework, and the carbon fibres reinforce it to make it stronger.

Carbon qualities >

Carbon-fibre polymers can be moulded into almost any shape. Changing the orientation of the carbon-fibre threads in a bike frame design can yield many different and desirable ride characteristics, from shock absorbency to stiffness. The degree of detail in the orientation of fibres, as well as types of resins used, affect the price of the end product enormously. It is possible to buy a basic carbon-fibre frame for a similar price to a good aluminium model, while the most sophisticated carbon creations can cost the same as a small car!

The carbon story >

Carbon-fibre bike frames first appeared in 1975. One of the early pioneers was the American company Exxon with its Graftek model. Exxon's early frames were very thin aluminium tubes wrapped in carbon, with the fibres aligned at a 10- to 15-degree angle to increase the stiffness. The resulting carbon-aluminium hybrid tubes were then sanded smooth. Like ALAN (see page 182), early construction was still quite basic and involved bonding the tubes together with aluminium lugs. This joining method proved to be an Achilles heel and many early models suffered joint failures.

The use of carbon-fibre in the cycling industry was advanced by French pedal pioneer Look. Its KG86 model of 1986 combined Kevlar with carbon fibre to increase the bike's strength and make it safer. The frame was constructed from Kevlar-reinforced carbon-fibre tubes bonded into thick lugs. Each frame was hand-made. The company's and the material's big bike breakthrough came when the American Greg LeMond won the 1986 Tour de France on a KG86.

Look went on to develop one of the first commercially available monocoque carbon-fibre frames, the KG196. This capitalized on carbon fibre's most valuable characteristic: it can be moulded into almost any shape. The KG196 was one of the most aerodynamic bikes of its day.

The French, through Look and fellow bike producer TVT, had a lead in carbon-fibre frame production. ALAN and other manufacturers followed by producing carbon-fibre bikes with the tubes glued into and joined by aluminium lugs. Meanwhile, a third French company, Vitus, who made aluminium frames similar to those produced by ALAN, also began producing a carbon frame. Both ALAN and TVT often sold their frames to other manufacturers who re-branded them.

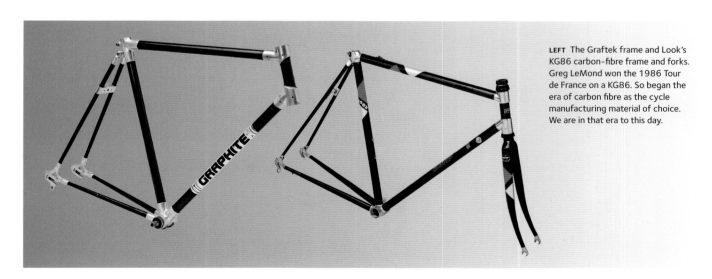

LEFT The Graftek frame and Look's KG86 carbon-fibre frame and forks. Greg LeMond won the 1986 Tour de France on a KG86. So began the era of carbon fibre as the cycle manufacturing material of choice. We are in that era to this day.

OCLV frames >

In the early 1990s, Bob Read, then technology director of American bike manufacturer Trek, saw how closed-mould manufacturing processes were used in the aerospace industry. He was convinced the same technique could be used to mass-produce carbon-fibre bike frames.

Trek invested heavily in the new technology and started using the process to manufacture its optimum compaction, low void (OCLV) frames. In 1992 Trek produced its 5500 frame, which weighed just 1.11kg (2.44lb) yet still exceeded aerospace standards. At the time, it was the lightest road frame in the world. Trek utilized the same production method to produce its 9800 and 9900 mountain bike (MTB) frames, weighing 1.29kg (2.84lb).

LEFT Trek's OCLV frame was a major step forward in the production of carbon-fibre bikes. The company received huge publicity when Lance Armstrong won the Tour de France from 1999 to 2005 on their bikes. (Armstrong was later stripped of his victories because of performance-enhancing drug use.) As a result, Trek became one of the world's leading bike manufacturers.

LEFT Cannondale experimented with the properties of carbon fibre in some of their mountain bikes, using careful orientation of fibres in the carbon weave to provide a degree of flexibility.

Precious Metal >

Of all the frame materials in common use today there is one that is light, stiff, strong, does not need painting and is virtually corrosion-free: titanium.

So why are all bikes now not made from this wonder material? Titanium is both extremely difficult to work with and it is very expensive. Consequently, it remains largely the domain of highly specialized custom builders with very discerning clients.

Some experimental frames were made with pure titanium in the 1960s, but the tubes had to be thick-walled to get the required stiffness; thus, much of the potential weight advantage was lost. To increase the stiffness, manufactures looked to the aerospace industry, where titanium had been alloyed with another exotic material, vanadium, and aluminium. The resulting materials were Ti-3Al-2.5V and Ti-6Al-4V; both are now the preferred alloys of choice for bicycle frames. Of the two, Ti-6Al-2V is the more difficult to machine and weld. Consequently, frames are more often made with Ti-3Al-2.5V tubes. The more cost effective Ti-6Al-4V is used for the dropouts and other solid parts.

Lightweight alternative >

Before the arrival of aluminium and carbon fibre, titanium was the only lightweight alternative to steel. So, for the racing elite, the difficulties of working with this exotic material were deemed worth the risk. In 1971, understanding of titanium was in its infancy and the vacuum brazing method used at the time produced joints that were prone to cracking.

OPPOSITE The Dolan ADX: a British titanium bike with carbon forks. This bike was designed for long-distance riding and sportive events. It has a longer head tube and shorter top tube to give the rider the option of a more upright riding position. The seat stays are curved, which adds to the comfort of the ride.

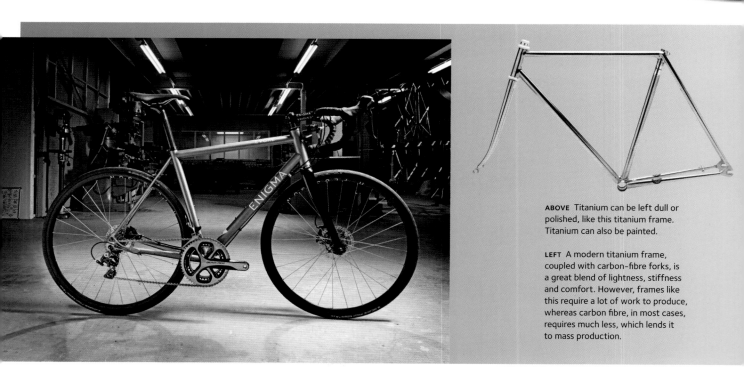

ABOVE Titanium can be left dull or polished, like this titanium frame. Titanium can also be painted.

LEFT A modern titanium frame, coupled with carbon-fibre forks, is a great blend of lightness, stiffness and comfort. However, frames like this require a lot of work to produce, whereas carbon fibre, in most cases, requires much less, which lends it to mass production.

Titanium reborn >

Cycle industry interest in titanium was reignited in the early 1990s when Japanese manufacturer Miyata brought out its bonded-titanium Elevation 8000 mountain bike. In the UK, industry giant Raleigh also introduced its versions of bonded-titanium road and mountain bikes as part of its Dyna-Tech range.

For a while, titanium was again vying to be the bike material of the future but its push for mainstream acceptance was quashed as consumers swayed towards carbon fibre. The trend away from metals was supported by the industry, aware of just how labour-intensive it was to work with titanium compared with carbon, which was more volume-manufacturing friendly.

Luis Ocaña >

The Speedwell Gear and Case Company made motorcycle frames for BSA's race division. In the 1960s the motorcycle giant asked Speedwell to make a motorbike frame out of titanium. Their calculations showed that a titanium frame would be 40 per cent lighter than a steel one, while retaining similar qualities. The resulting frame was indeed less than half the weight but it cost a staggering £4,000. A steel race frame of the day had a price tag of just £40. Unsurprising, the project was abandoned.

Having developed its expertise, Speedwell turned its attention to bicycles, whose frames were more sophisticated than those of motorcycles. Yorkshire wholesaler Ron Kitching got involved with marketing and, through his connections in the French cycle trade, convinced the Bic pro team to let its star rider, Luis Ocaña, try one.

In the 1973 Dauphine Libere, Ocaña used a Speedwell Titalite frame on all the mountain stages and won the event handsomely.

Having gained confidence in the new frame, he continued to use the Titalite through the mountains of that year's Tour de France, which he won (in the absence of Eddy Merckx) by a massive 15 minute margin. The Titalite could not have had a better launch but sadly it was a case of right product, wrong time. Britain was in the midst of an economic downturn and the Titalite was extremely expensive. Although sales started strongly, sales dropped away and Speedwell folded in 1977.

BELOW Luis Ocaña (second) follows fellow Spaniard José Manuel Fuente during a mountain stage of the 1973 Tour de France. The Aérospatiale plant in Toulouse drilled holes in and shaved metal from the Campagnolo components of Ocaña's lightweight titanium bike, producing a machine that weighed 7.8kg – 2kg lighter than other race bikes at the time.

Chapter 10
Body Language >

The author setting a new world hour record of 56.375km/h (35.029mph) in 1996 using what is probably the most aerodynamic position ever on an upright bike; the Superman position. The position was developed by Graeme Obree, who used the position to win the 1995 World pursuit title.

The First Tri-Bars >

The idea of fitting an attachment to dropped handlebars to give a more aerodynamic riding position came first from the world of ultra-distance cycling and was developed by somebody with a skiing background.

Ultra-distance racers seek comfort as well as speed. In 1984, cyclist Jim Elliott rode to fourth place in the 4,800km (3,000 mile) Race Across America (RAAM) on a bike with centrally placed arm rests on its handlebars. That year's race winner, Pete Penseyres, adapted Elliott's idea and, in 1986, competed with a handlebar attachment that allowed him to ride with his forearms supported and his hands extended centrally in front of him. The handlebars supported Penseyres' upper body, saving him energy, but the flat, frontal position of his forearms took them out of the airflow.

The first tri-bars >
Boone Lennon was the US ski team coach from 1984 until 1986, and Lennon's ski experience led him to think about the way cyclists rode their bikes. He believed that bringing cyclists' arms more in line with their body centre-line as well as flattening their forearms, like downhill skiers do in a tuck position, would make cyclists more aerodynamic, and therefore faster. So he developed the first clip-on tri-bars.

In 1987, one year after Penseyres's winning ride in the RAAM, Boone Lennon patented a design for clip-on handlebars. His invention, clearly influenced by the RAAM legend's pioneering idea, was simple: an aluminium tube bent into a U-shape and clamped to the top of a bike's existing handlebars. Elbow pads were added to the handlebar tops; the rider crouched down, rested his or her elbows on the pads and held the forward part of the U-shaped tube with his or her hands.

This new position brought the rider's arms within the silhouette of the body, thereby reducing the frontal area dramatically. Additionally, the action of bringing in the elbows also rounded off the rider's shoulders, smoothing the flow over their back. A final bonus came not from aerodynamics but from ergonomics. A cyclist holding tri-bars in a tuck position holds the front part of the tri-bars while their arms rest on pads, so the weight of their torso is supported by their skeleton rather than their arm muscles. This new braced stance allowed the cyclist to roll forwards, reducing the frontal area further while requiring less energy to maintain.

Lennon's clip-on handlebars were produced by the ski equipment manufacturer Scott. As early adopters of the bars came from the world of triathlon, they became known as tri-bars, the name they still hold today.

RIGHT The cycling world first became aware of tri-bars in 1989, when the American Greg LeMond won the Tour de France during the final time trial in the streets of Paris by just 8 seconds, still the smallest-ever victory margin. No piece of cycling equipment has ever debuted in a more dramatic fashion.

In 1989, the American professional cyclist Greg LeMond tried Lennon's bars and was so impressed that he decided to use them for the final stage in that year's Tour de France, a time trial. LeMond won the race, turning a 58-second deficit into an 8 second advantage and so snatching the overall victory from French star Laurent Fignon. He became the first American winner of the race. His performance, undoubtedly aided by his adoption of this new technology, made tri-bars the new standard for time trials in cycling.

Big step >

This piece of equipment was nothing short of revolutionary and changed time-trial bikes for ever. Many variations of Lennon's basic design followed, with other manufacturers keen to capitalize on the new trend.

Athletes and coaches experimented with new positions to try to maximize the possibilities that the braced position allowed. Saddles were pushed further and further forward as the riders' bars got lower and lower. To accommodate all of these new changes, the basic geometry of the bike had to change too.

Seat tube angles became much steeper, top tubes shorter and gear shifters were adapted to fit on to the tri-bars so riders did not have to move from their tucked pose to change gear.

Integrated tri-bars >

For a long time, the tri-bar was treated as a mere accessory, something that would be attached to standard dropped or bullhorn handlebars with a few bolts. Then, in 1991, realizing that tri-bars were here to stay, the French bicycle product company Corima produced the first integrated tri-bar, incorporating both outboard position, where the brakes are located, and the new TT bars into one product. Later on, even the brake levers were incorporated.

Although the basic design remains unchanged today, some manufacturers are starting to make the bars and frame into one flowing design. The resulting products are aerodynamically superior and aesthetically pleasing.

ABOVE As well as being very aerodynamic themselves, modern integrated tri-bars are designed so that a rider is able to get into an aerodynamic riding position. They are also often designed by, or with, bike manufacturers, so that they work aerodynamically with the rest of their bike.

LEFT/FAR LEFT The search for an aerodynamic optimum can be taken a stage further by designing a single bike for a single rider, as in the case of the author's Olympic gold medal-winning Lotus bike, ridden at the 1992 Barcelona Olympics. The bike's carbon-fibre handlebar and tri-bar unit was made by creating a unique mould. But even a bike like this has to be powered by a strong athlete to achieve the best results.

Aerodynamic Road-Race Handlebars >

After their success with the time-trial handlebar (see pages 190–91), Greg LeMond, Boone Lennon and the Scott company carried on experimenting – this time with road handlebars, where they applied some very lateral thinking.

Scott dropped handlebars were based on standard dropped handlebars but had an extra horizontal bar at the bottom on each side. The idea was that rider could hold this lower bar when descending, supposedly allowing them to crouch lower and reduce their total frontal area further. Unlike with tri-bars, the rider was required to hold this new position without elbow support, making it uncomfortable, less stable and only debatably more aerodynamic. The product was short-lived.

Scott's next offering was even more radical and even less practical. It was a mini tri-bar extension that could be added to the low bar, supposedly to bring the rider's arms closer. However, without elbow support it was impractical to hold and even more unstable than their previous invention.

Tri-bars for road-race bikes >
The Italian company Cinelli's Spinaci clip-on handlebars from the early 1990s were a much better solution. Designed for road bike use, the Spinaci bar was effectively a miniature tri-bar bolted on to the centre of road bars. While not allowing the rider to crouch low, they crucially enabled the rider to rest their forearms on the handlebars, giving vital stability and a degree of comfort. For the short time these bars were allowed in competition, they were extremely popular, both for uphill time trials and road races. In 1997, they were deemed to be a safety hazard by the UCI and banned from competition.

RIGHT Greg LeMond, on his way to winning the 1990 Tour de France, was chief test pilot for many of Boone Lennon's handlebar designs that were branded by Scott. The clip-on tri-bar pictured here is a neater version of the one LeMond used to win the 1989 Tour de France.

LEFT Tri-bars are not only aerodynamic; they also help support a cyclist's upper body, removing the strain of doing so from the arms. The handlebar adaptation used by this rider shows him resting his arms rather than using them for support.

LEFT Spinaci clip-on handlebars were outlawed by the UCI in 1997, but road racers still rest their forearms on the top of standard dropped handlebars in an effort to improve their aerodynamics. Racers often do this when riding alone on the flat in an effort to ride faster, as Germany's Tony Martin is doing here.

ABOVE The author using the Superman riding position during his successful world hour record attempt in 1996. With arms placed almost straight out in front, the drag is dramatically reduced. The lowered and stretched position of the upper body reduces frontal area considerably and smooths out the airflow over the entire torso.

Evolution of Aerodynamic Clothing >

If you look closely at the world's top time triallists, you will noticed that their clothing is not simply body hugging and wrinkle free, but is also made up of different textures and embossed patterns. This is not a fashion statement but a sophisticated way of managing the airflow over different body parts. Despite different teams having different clothing sponsors – which contribute considerable sums of money to each team's budget – the effect of clothing on performance is so enormous that riders can often be seen wearing garments made by the same manufacturer, with logos and badges carefully removed.

The Belgian manufacturer Bioracer was one of the first companies to explore and exploit clothing aerodynamics on a commercial scale, doing so since the start of the 21st century. Its research and development process showed that a mixture of smooth and rough materials on different body parts helped manage airflow.

Its individual speedwear range is custom-made for each cyclist, not just for fit but also for the specific speed range in which he or she will use that garment. Bioracer's research indicated that a material combination that worked well at 50–55km/h (31–34mph) – a speed zone applicable to the world's best in a flat time trial – does not work well for a track sprinter in the 60–70km/h (37–44mph) range.

Each speedwear skinsuit costs many hundreds of pounds and lasts for only a handful of races before it loses shape. This may seem extravagant, but at the 2012 Olympic Games 12 of the cycling medallists wore Bioracer suits.

The technology has now been made more affordable, and so performance-enhancing garments are being bought by sporting enthusiasts around the world.

RIGHT/BELOW The evolution of aerodynamic clothing design. Creases are clearly visible in the early 21st century aerodynamic clothing worn by the American rider below, whereas the latest in aerodynamic design worn by the 2011–2013 world time trial champion Tony Martin of Germany (right), is almost entirely smooth, allowing air to flow smoothly over it.

The kite suit >

It is not just motor sports that push both technology and the regulations to the limit in an effort to gain an advantage over rivals. In 2010, Pearl Izumi, a Japanese manufacturer, developed a skinsuit with material 'wings' that bridged the gap between arm and torso, thereby smoothing the airflow as it passed over the body. Theoretically, these wings also increased a rider's side area and acted as a sail in crosswinds, effectively pushing them along.

The suit was purportedly worth 20 watts of extra power, which is about a 5 per cent gain for a good professional rider. After examining the garment, the UCI decreed that it did not follow the form of the body and constituted a fairing. The suit was duly outlawed.

RIGHT By using material to bridge the gap between the rider's upper arms and torso, the suit smooths the air flowing over the transitions between arm and torso. The extra material also forms 'wings' that can catch a tailwind, adding propulsion.

Aerodynamic Helmets >

Protective headgear is now mandatory in all international competitions, so it is understandable that much time and effort has been invested in improving the aerodynamics of cycling headgear.

In the early 1990s, protective foam cycling helmets became popular, adding 5cm (2in) to the rider's profile, along with a significant amount of drag. In 2003, after several failed attempts, the protective helmet became mandatory for international cycle racing, making investment in helmet development a necessity.

Most riders, even the very best, cannot tuck their head within the 'silhouette' of their body; instead, it sticks up and accounts for between 10 and 20 per cent of their profile. As it is not possible to reduce this frontal area, the best way to deal with it is to change the shape of the helmet. Removing the turbulence-inducing holes of a standard helmet and elongating the tail can have a significant effect on reducing overall drag.

Aero evolution >

The first 'aero' helmets were just fabric covers for the old-fashioned leather ribbed helmets, as worn by Francesco Moser in his successful 1984 attempt to beat Eddy Merckx's hour record in Mexico City.

Fascinated with aerodynamics after his successful attempt, Moser continued researching and was one of the earliest pro cyclists to frequent a wind tunnel. More sophisticated shapes followed and his ribbed helmet was discarded in favour of a smooth thin plastic shell.

The topic of aerodynamic helmets became fashionable during the early 1980s when several new designs appeared. Some were minimalist plastic skull caps, while others boasted teardrop-shaped tails.

The American Olympic team were among the first to wear what were effectively fibreglass head fairings for the pursuit events in the 1984 Games. In 1985, Bernard Hinault used a Cinelli AeroLite plastic helmet in the time-trial stages of that year's Tour de France. However, none of these early creations had any meaningful protective capacity – a fact that the UCI did not address for more than a decade.

Even for hazardous track sprint events, where crashes were common, aerodynamics took precedent over safety. The all-dominant East German sprinters of the early 1980s all used LAS plastic helmets with only thin, comfort padding.

RIGHT/OPPOSITE Cinelli was one of the first manufacturers of aerodynamic helmets. The helmet worn by Bernard Hinault in the 1985 Tour de France Prologue time trial smoothed the air flowing over his head, but it had few protective qualities. The victorious East German rider opposite was also wearing an aerodynamic helmet that gave little or no head protection. From 2003, any cycle helmet worn in competition had to pass rigid safety standards to ensure sufficient head protection was provided.

Helmet timeline >

1989

Greg LeMond's Giro Aerohead

Giro, an American company, was at the forefront of early protective helmet development. Its founder, Jim Gentes, was particularly interested in aerodynamics and oversaw the creation of the Aerohead, a fully protective foam helmet with a teardrop shape. In an attempt to smooth airflow over the sides of the head, Giro's aero creation even covered the ears. It was the headgear used by Greg LeMond to win the 1989 Tour de France final time trial, where he won the race overall by the narrowest ever margin of just eight seconds.

1994

My custom Giro 'bug'

In my debut Tour de France, I used a minimalist head fairing with built-in visor made by Giro to win the opening prologue time trial around the streets of Lille. In doing so I set a record speed for a Tour de France time trial – a mark that still stands today. The short tail design has stood the test of time and is remarkably similar to the Kask model used by Sir Bradley Wiggins some 18 years later to take overall victory in the 2012 Tour de France. Many manufacturers followed the trail blazed by Giro and addressed the aerodynamic challenge in new and innovative ways.

1995

Miguel Indurain's Rudy Project Sweeto

Indurain's helmet, with its space-age design, was one of the first helmets that smoothed out air flowing over a rider's face.

2000

David Millar's Giro Rev IV

This is another full-face aero helmet design.

Sir Bradley Wiggins on his way to winning the time-trial gold medal at the 2012 London Olympics. He's riding a custom-made time-trial bike supplied by British Cycling – the product of years of painstaking research and collaboration with aerodynamic experts, not just from cycling but from the aerospace and motor-sport industries.

Super-Bike v Washing Machine >

The battle of the bikes, between Old Faithful and the Lotus 108, was described by some as cycling's David and Goliath, as gut feel and intuition pitted against hard science, a one-man band taking on a top car manufacturer with space-age materials and a racing heritage. In reality, it was two committed designers – athlete Graeme Obree and inventor Mike Burrows – both trying to achieve the same objective but with two very different approaches.

The Lotus super-bike >

When I took the gold medal in the track pursuit event at the 1992 Barcelona Olympics, it was the first Olympic Gold in cycling for Great Britain for 72 years and took everyone, including me, by surprise. Despite having broken the world record in training, I still believed that winning was what other countries did (such as people on the television), not unemployed carpenters from Liverpool. The UK press was completely unprepared so, unsurprisingly, as they scrabbled for information in this unfamiliar environment, their focus was drawn to the bike that had been supplied by the car manufacturer Lotus. It was like nothing people had ever seen before.

In the early 1980s, bike designer, inventor and amateur aerodynamicist Mike Burrows had begun to use carbon fibre in some of his creations. He was one of the very first to recognize its full potential. Carbon fibre was fairly new to cycling (see pages 184–5) and manufactures were not quite sure how to use it, so they stayed in their comfort zone and rolled the wonder material into conventional tube shapes. Burrows, however, took fuller advantage of carbon fibre's most valuable property: its ability to be shaped. Instead of sticking tubes together, he made a wholly new shape – a frame manufactured in a single form, known as a monocoque. Manufacturing in this way was hard but the end result was not only a strong frame but also a shape that addressed the biggest force the bike needed to overcome – air resistance. Burrows filed a patent for a monocoque bike frame in 1982, and by 1984 he had produced his first rideable monocoque.

The Norfolk engineer questioned every aspect of bike design, including the traditional way of making the forks, the part of the bike that hits the air first. Burrows reasoned that if fork blades caused drag, why have two of them? The inherent strength of carbon fibre allowed him to use a single, wing-shaped blade with the front wheel axle bonded into it; the wheel was then slid on to this from the side. The same unique method of joining wheel to frame was used at the rear of the bike. The fixed rear axle meant adjusting chain tension had to be achieved by the novel method of rotating the bottom bracket, itself mounted in a concentric circle arrangement. This had the unfortunate side-effect of requiring saddle height to be adjusted every time the gear ratio was changed! It was not an easy machine to work with.

Burrows's original machine, a time-trial bike using the same temperamental arrangement as the final track bike, was also a thing of aerodynamic beauty but not everyone agreed that the technical challenges were worth the trouble or that aerodynamics were the future. Burrows tried to sell the idea to the British cycle industry. They were not interested, but I was.

Rudy Thoman was a test driver for Lotus cars and a friend of Burrows. On seeing his design, Thoman saw an opportunity to get the Lotus name into the Olympics. The car company liked the idea, as well as the engineering and sporting challenge the project would present. The only thing Lotus needed was someone to ride the new bike. Burrows was aware of my pursuit background and believed that a track version of his bike might offer the edge I needed to close the gap on Olympic gold.

In the winter of 1991–2, Burrows invited my coach Peter Keen and myself to a wind tunnel session at the Motor Industry Research Association (MIRA), just outside Birmingham. The results convinced Keen, myself and the British Team that this was the machine Great Britain should be using for the Barcelona Olympic Games.

In 1990, although previously banned, the UCI decided to relax its rules on technology and allow monocoque frames in competition. It was a decision they would later regret but for now it meant that the Lotus super-bike project could go ahead.

RIGHT/BELOW The Lotus super-bike stood bike design on its head and opened up a world of possibilities. The sweeping lines look fast in side view, but the true aerodynamic advantage can only be appreciated from the front, as the bike has an incredibly narrow front profile. The integrated seat post improves airflow but means the bike will only fit the rider it was built for [1]. Clipless pedals are biomechanically efficient and improve airflow [2]. Custom-made handlebars allow the author to crouch low on the bike [3]. As the rider is the largest part of the bike/rider package, the advantage of the aerodynamic position achieved was enormous.

OPPOSITE The author catches his German opponents in the final of the 4000-metre (4374-yd) individual pursuit to win the gold medal at the Barcelona Olympics in 1992.

1

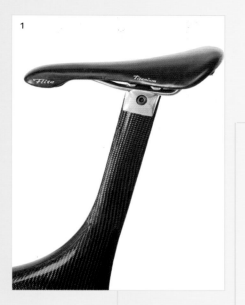

2

3

Old Faithful >

A few years earlier, Graeme Obree had approached the aerodynamic challenge from a wholly different angle, when he created the bike that came to be known as Old Faithful. It might have been rudimentary in construction but it would be used by Obree in a successful bid to break the most prestigious of all cycling records – the hour record held by Francesco Moser since 1984.

In his design, Obree focused not on the aerodynamics of the bike but on those of the rider – his rudimentary reasoning being that if he could reduce the size of the rider this would have a big impact on speed. He was correct. Rather than the standard triathlon bars, Old Faithful had a high, flat 'crossbar'

with mountain bike grips at the ends. Holding these, Obree crouched forward and rested his shoulders on his hands, thereby effectively hiding his arms completely from the airflow. He improved the cross-sections of his handlebar by squashing it in a vice until it was oval!

It might have been crude but it gave Obree an incredibly small frontal area. In an attempt to reduce his profile even further, he created a super-narrow bottom bracket axle, for which he needed an equally narrow bearing. This he famously scavenged from an old washing machine. Although it probably did not achieve its primary goal (as it did not reduce the size of his legs, just brought them closer to each other) it almost certainly improved his mechanical efficiency.

To accommodate the legs being so close together, Obree's home-made steel frame had a single sloping crosstube that continued down into the chain stays, thereby doing away with the need for a top tube.

On 16 July 1993 in Norway, Obree failed at his first attempt at the hour record. Incredibly, he went back the very next day and succeeded, setting a new distance of 51.596km (32.06 miles). My attempt was scheduled to take place much further south, in Bordeaux, France, a few days later. On 23 July 1993, I broke Obree's new record with a distance of 52.270km (32.48 miles), using a carbon-fibre frame supplied by French manufacturer Corima.

OPPOSITE/BELOW Old Faithful was not a very aerodynamic bike; its design focused on giving its rider, Graeme Obree, the most aerodynamic position. These pictures show how low his upper body is in relation to the track, how his head is lower than his back, and how his helmet helps to smooth the airflow between the two.

LEFT The front view shows how Old Faithful's design removed Obree's arms almost totally from the airflow as they were tucked behind his shoulders. His hands are also well inside the overall frontal area of his body, so they don't create an extra drag source.

Superman's Solution >

Sadly, after breaking the hour record on 17 July 1993, Graeme Obree was not immediately recognized as the amazing athlete he was. Instead, his success was wholly attributed to Old Faithful, built effectively from scrap, which he used to achieve his various feats (see page 200). Only a formidable athlete could break the blue ribbon record of cycling. Many also thought – as they had 60 years earlier when Francis Faure broke the hour record on a recumbent cycle – that what Obree was doing was not cycling at all.

During the winter of 1993–4 I tested one of Obree's bikes and confirmed that, although the machine was ungainly, the concept was effective but not definitively so. Nevertheless, the UCI decided that this was not an aesthetic they wanted to see prosper in the sport so set about creating a succession of regulations that would make it impossible for Obree to ride Old Faithful in international competition.

On 27 April 1994, while the UCI continued to work on the rules, Obree broke the hour record once more, achieving a new mark of 52.713km (32.75 miles) – this time on the same track at Velodrome du Lac, Bordeaux, that I had used to break the record the previous year. The UCI's new regulations came in soon after.

At the world track championships on the Isles of Scilly that August, Obree frantically adapted his machine in order to comply with the new rules. Thinking that he had succeeded, he woke on the morning of the qualifying round to find that another sentence had been introduced to the rules overnight. This stated that there had to be 'a certain distance between a rider's chest and handlebars' and 'daylight' must be visible between the chest and arms. This clarification thwarted Obree's attempt to compete. Officials watched Obree ride in qualifying, crouching low to check where his chest was in relation to his handlebars. Obree could not comply with the new rule and was disqualified during the opening round.

RIGHT Graeme Obree and the second ground-breaking riding position he created: the Superman. This position was even more aerodynamic than the one he used on Old Faithful. Locking his arms almost straight out in front of him allowed Obree to breathe freely too.

The main components of his new machine – frame and forks – remained unchanged, but the flattened crossbar was replaced with a long stem and extra-long tri-bar set almost as high as his saddle. The resulting stretched-out position looked remarkably like Superman.

Obree's response >

Obree's response was not to give in. Instead, he came up with yet another radical solution, based on the same strategy: improve the aerodynamics of the body not the machine. The main components of his new machine – frame and forks – remained unchanged, but the flattened crossbar was replaced with a long stem and extra-long tri-bar set almost as high as his saddle. The resulting stretched-out position looked remarkably like Superman, so unsurprisingly this is what Obree's new set-up came to be nicknamed.

Ironically, this new UCI-legal position was even more aerodynamic than his original idea. His upper body was locked in a flat position, with arms outstretched. This was remarkably efficient and allowed excellent lung expansion. Not only was the frontal area reduced by placing the arms in line with the body, the airflow over the shoulders and torso was also much less turbulent, reducing the drag forces even further.

Superman flies >

Astride his Superman bike, Graeme Obree won his second world pursuit title in 1995. Much to my delight when I trialled it in 1996, the design was probably even better suited to my body shape then it was to its designer's. In 1996, I used Obree's pioneering work to win the world pursuit championships with a new record for the 4,000-m (4,374-yd) pursuit distance of 4 minutes 11.114 seconds. A few weeks later, I again used the Superman position to break the world hour record posting a distance of 56.375km (35.03 miles). The mark would stand for 15 years. The regulations were again changed, outlawing Obree's new strategy, so the hour record mark I set using the position has never been beaten.

BELOW The Superman position suited the author's body shape even more than it did Obree's. Using it, he set a new world hour record at the Manchester Velodrome in 1996. However, the position was disallowed soon after the event and nobody has since gone further.

The Athlete's Hour

Most top track riders who specialized in the pursuit were forced to copy Graeme Obree's Superman position (see pages 204–5) because its advantage was so great. A brave few even tried out the position for road time trials – an environment it definitely was not suitable for. The UCI stance was predictable; it looked fundamentally different to the last 100 years and had more in common with human-powered vehicle (HPV) racing. Thus, they did not want to see the Superman position anywhere.

The UCI had modernized when they accepted triathlon bars (see pages 190–1), but the extremely specialized Superman position was more than it was prepared to tolerate. Having allowed a track world championship and two new records to be set using the highly advantageous position, the UCI outlawed it as part of a raft of technology-controlling measures brought in after the 1996 Olympic Games.

The new rules effectively took cycling back a decade: from that point on, bikes would be just a collection of tubes. This winding back of the clock left the UCI with a dilemma. It had allowed new records to be set using advantages now not available to future riders. Its immediate response was to let the records stand and ignore them.

Private project goes public >

It was four years later, on hearing about my decision to set a new mark – a wholly private project – using a bike of similar technology that had been available to Eddy Merckx back in 1972, that the UCI saw an opportunity to reconsider its stance.

My exercise was actually conceived in Bordeaux in 1992, while tackling the hour record for the first time, and it was rekindled in 2000 as a poetic swan song. Having been involved with high technology throughout my career, I felt that it was a nice way to finish, to 'clean up before I left' by setting a mark that was purely about athletes. By using a bike with round tubes and spoked wheels, the bike was effectively removed from the equation, allowing cycling fans to compare riders from 30 years ago to the current generation.

RIGHT Since the UCI banned the Superman position shortly after the author's 1996 world hour record of 56.3759km/h (35.0303mph), that record is now referred to as the Best Human Effort. Following the ban, the UCI reinstated Eddy Merckx's 1972 record as the official hour record. In 2000, the author beat Merckx's record riding in a standard cycling position on a similar type of bike to the one Merckx had used. This new record became known as 'the Athlete's Hour'.

> Having been involved with high technology throughout my career, I felt that it was a nice way to finish, to 'clean up before I left' by setting a mark that was purely about athletes. By using a bike with round tubes and spoked wheels, the bike was effectively removed from the equation.

With just weeks to go, the UCI decided to adopt the athletes' hour record (as they christened it), effectively joining up Eddy Merckx's Mexico City record set in 1972 with the present day. Using this set of rules, even the marks set by Fausto Coppi in 1942 and Jacques Anquetil in 1956 would become directly comparable with riders far into the future. The hour record became known as the athlete's hour and my Superman-enhanced records were left on the shelf to gather dust.

In October 2000, I managed to beat Merckx's mark by just 10m (33ft), recording a distance 49.441km (30.72 miles) in the hour. It was the last thing I ever did as a professional bike rider.

A Czech rider, Ondrej Sosenka, improved the mark, setting a new distance of 49.7km (30.88 miles) in 2005. From there, the hour record went into a period of dormancy – something it has done several times over the years.

New interest in the Hour >

Interest looked to have been rekindled in 2013 when several high-profile cyclists were linked to pending attempts on the athlete's hour record. This revival of interest jogged the UCI – now under new management – to re-examine the rules surrounding the record. After 14 years, it decided that it was time for a change and, in May 2014, it announced that the athlete's hour would be scrapped in favour of letting athletes go back to using equipment permitted under the rules of the day. This realigned it with the ethos of the past and cut the direct links to the Merckx era. At the time of writing, several riders have picked up the hour record baton.

The record that the UCI decided they had to beat was the furthest athlete's hour figure of 49.700 kilometres set by the Czech, Ondrej Sosenka in 2005. The first to do it was Jens Voigt of Germany in September 2014 with a distance of 51.110 kilometres. Then in October 2014 Mathias Brandle of Denmark went further with 51.855, and in February 2015 the Australian, Rohan Dennis set a new men's world hour record of 52.491 kilometres in Switzerland.

LEFT/ABOVE Eddy Merckx setting his world hour record in 1972 and the bike he did it on. The bike is displayed in a glass case on the platform of the Eddy Merckx station, which is part of the Brussels Metro on Line Five to the west of the city.

Power Meters

The arrival of heart-rate monitors in the early 1980s (see pages 180–1) gave cyclists and their coaches a practical way to objectively quantify effort. Training was defined by heart-rate bands rather than by subjective descriptions. Although practical and cost-effective, heart-rate monitors had their limits. An athlete's pulse rate is affected by factors such as climatic conditions and fatigue, so even with the same perceived effort, it can vary significantly. As pulse reflects physical state (fresh or tired) it is useful, but as a way of dictating workload it is highly inaccurate. A more useful way to measure and quantify training intensity is power.

Static bikes called ergometers had been used since the 1960s and measured an athlete's power output, expressed in watts. These ergometers were and still are used by coaches and sports scientists to assess a rider's physical capabilities. It gradually became self-evident that the ability to measure power while actually out on the roads in real time would be highly desirable. If this could be achieved, training efforts could be proscribed, executed and analyzed with much more accuracy.

Pioneer >
The first accurate and reliable bike-mounted power meter was made by German engineer Ullrich Schoberer. Built into the right-hand chain-set, Schoberer's meter used a number of strain gauges to measure the torque forces in the crank and then converted this measurement into watts, which were displayed in real time on a handlebar-mounted display. The rider could use the information to monitor his or her effort with incredible accuracy. After the session, the information could be downloaded for analysis or sharing with a coach. In 1986, Schoberer formed a company, Schoberer Rad Messtechnik (SRM), and made his creation available to the public. The SRM system still sets the standard for power-measuring devices today but it is coming under pressure from several new companies, eager to capitalize on the demand from cyclists and triathletes for accurate training quantification.

Re-inventing the wheel >
The power generated by a rider is transferred via the pedals and crank, through the chain to the rear wheel on to the road. Although Schoberer's device used strain gauges located inside the chain-set, other designers have chosen different ways to measure rider-generated forces.

Like the SRM system, the Power Tap system, developed during the 1990s, also uses strain gauges but these are mounted on a torque tube housed inside a custom rear hub. The information is conveyed to the rider in the same way, as watts on a bar-mounted

LEFT Market pioneer SRM set the standard for power meters. Wireless options are available today, built into a wide variety of chain-sets.

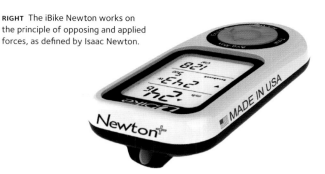

LEFT/BELOW A Power Tap rear hub built into a wheel, and the rear and front hubs shown on their own. Information from the hub's can is picked up by a variety of Ant+ devices mounted on the handlebars.

RIGHT The iBike Newton works on the principle of opposing and applied forces, as defined by Isaac Newton.

display. Unlike the SRM-equipped cranks, which are permanently attached to the bike, the Power Tap wheel can be removed from a bike or easily transferred between machines.

Pedal power >

Seeking to simplify the system even further, Garmin worked with the pedal manufacturer Look for several years to develop its Vector power-measuring pedals, prior to its general release in 2013. Force-sensors placed in each pedal measured the slight deflection in the pedal axle. This information not only provided

total power output but also accurately recorded how much of that total each of the rider's legs contributes. This feedback could be used to help a cyclist overcome asymmetry in their development.

The power of physics >

The iBike Newton power meter, which entered the market after 2010, utilizes Newton's third law of motion which states that opposing forces equal applied forces. Instead of measuring torque data like most other power meters, the handlebar-mounted Newton calculates its data based on acceleration, opposing air pressure and degree of elevation as the cyclist rides. Speed is also recorded using a simple remote sensor. To provide the opposing forces for the equation, bike-plus-rider weight is entered into the device at set-up, along with tyre size, road surface type, rider height and ride position. The Newton then uses all these inputs to calculate the rider's coefficient of drag (CdA) and bike coefficient of rolling resistance (Crr). In addition, when riding, the device uses the same information, along with speed, to predict power output.

Stages racing >

The Stages power meter, which was incorporated in Shimano, SRAM, FSA and Cannodale cranks in 2014, uses a custom-designed strain gauge to measure power. Stages claim that this method is both lighter than other power metres and does not compromise chain-set stiffness. Shimano also claimed that this method was preferable, after research showed that chain torque was affecting results in other devices that measured power through the right-side chain-set or rear hub.

BELOW GPS manufacturer Garmin worked with Look to develop the Vector power measuring pedal. Power information recorded at the pedal is transmitted to a metal transmitter that fits between the pedals and the bike's cranks. Information from them is received by and displayed on a handlebar-mounted power-enabled GPS for real-time measurement.

BELOW RIGHT One of the latest power meters to be developed, Stages uses a strain gauge mounted in the left-hand crank.

Triathlon Bikes >

The bike leg of an elite Olympic triathlon resembles a mini road race, with drafting allowed, so the competitors use road bikes. The same rules apply for the triathlon World Cup series and the Olympic-distance world championships. The Olympic event, however, does not reflect the majority of triathlons, most of which forbid drafting. For this, a time-trial type machine, equipped with aero bars and deep section wheels, is more the norm.

3 Saddles like this are designed specifically for use with tri-bars. They support the area of the pelvic bone structure that comes into contact with a saddle as a rider crouches and moves forwards, to hold the tri-bar extensions in an aerodynamic tuck.

4 A recess in the aerodynamically shaped seat tube allows the rear wheel to be sheltered more effectively from the wind, thereby reducing drag.

1/2 Details of the rear brake attached underneath the bike, where it is more hidden from the airflow than it would be in the standard position. The wide range of sprocket sizes means a rider could tackle a variety of terrain on this bike.

5 Placing gear shifters on the ends of tri-bar extensions means shifts can be made while riding in the most aerodynamic position. The control cables on this bike run inside the handlebars to keep them out of the airflow for as long as possible.

6 The front brake is hidden in a recess located inside the forks, so that it doesn't interfere with air flowing over the forks and cause turbulence, which increases drag.

7 The inner sides of the forks on this bike are flat, which reduces turbulence caused by the interaction of air coming off the front wheels and the forks.

Design Shift >

Although it is the model that he is most famous for, the Lotus super-bike (see pages 200–1) was just one of Mike Burrows's designs. His engineering skills and imaginative approach have been used on projects as varied as human-power vehicles to utilitarian cargo-carrying bikes. In fact, it was one of Burrows's more conventional designs that has had the longest-lasting effect on the bicycle world.

Working with bike manufacturer Giant, Burrows designed the compact frame. It was a concept that would subtly change the look of road-racing bikes forever. Instead of a conventional horizontal top tube, Burrows sloped this tube down towards the seat tube, making the main triangle smaller, stiffer and lighter. Giant called Burrows's creation their Total Compact Road (TCR) design.

An added advantage of minimizing the triangle size and lowering the top tube was the increase in stand-over height. (This is the distance between the rider's crotch and the top tube when they are stood with their feet on the floor, either side of the bike.) This meant each size of bike fitted a wider range of people. The TCR heralded a move away from custom-made frames towards off-the-peg frame sizes, even for expensive models.

Devil in the detail >
In order for the top tube to slope backwards, an extra-long seat post was required. Rather than use a round seat tube, Burrows supplied each of his frames with a carbon post that had an aerodynamic profile.

RIGHT A modern example of Mike Burrows's compact bike frame. The seat tube on this model continues past the top tube to form an integrated seat tube and seat post, or seat-mast. Adjusting an individual rider's leg length is made by cutting the top of the seat tube to suit, then fitting a metal collar with the saddle bolted to it over the carbon-fibre end.

RIGHT Close-up of an integrated seat post and seat tube. This is a neat, light alternative to the traditional seat post, part of which sits hidden inside the frame and is extra weight to carry.

Mike Burrows >

Burrows did not restrict his designs to wheeled projects. He was an accomplished model maker too, and represented Great Britain in the world gliding championship for 1.8-m (6-ft) wingspan models. He took up cycling while working as an engineer, where his inventive mind started to work on how bikes could be improved.

Of all his design successes, it is for his work on the Lotus super-bike that he will almost certainly be remembered – it is certainly what I remember Burrows for! I am sure that, had the UCI regulations not been changed to outlaw monocoques (see pages 204–5), his creation would still be in use today. Although less dramatic in appearance, his idea of a compact frame, which has a sloped-back top tube making it smaller, lighter and stiffer than a standard geometry road frame, is still in vogue today. The style was originally championed by Giant but is now popular with many other manufacturers.

Burrows still designs bikes but now focuses more on the off-shoot area of human-powered vehicles (HPV) (see pages 82–3), where his imagination has free rein. He also believes that freight-carrying bikes have a real future as a local sustainable transport option – a point on which I firmly agree. Inspired by the long bikes that the Viet Cong used to transport goods along the Ho Chi Minh trail, his 8Freight bike is able to carry loads from shopping to children and exemplifies the flexibility of his engineering genius.

BELOW/RIGHT Burrows' version of the long bike used by the Viet Cong, a sustainable and adaptable vehicle that is suited to a number of load-carrying functions.

Exploding Bikes >

The story of the bicycle is littered with new designs that utilize exotic materials and processes in their construction. Over the past 200 years, nearly everything has been tried, from wood and iron to carbon fibre and titanium – even paper. Indeed, 200 years from now the search for better ways to make bicycles will almost certainly still be going strong.

The magnesium bike >

Perhaps one of the wildest concepts was dreamed up by Ford car designer Frank Kirk, who had experience in magnesium casting. Kirk believed that because magnesium was abundant and could even be produced by electrolysis from sea water, it could be a good material for bike frames. He postulated that there should be enough magnesium in 1cu m (35cu ft) of sea water to make a bike frame. He also reasoned that, by using injection moulding technology, it should be possible to produce a frame in as little as eight seconds.

Initial pilot models seemed to support Kirk's theory. His cast magnesium bikes were so strong that the prototypes were left almost unmarked after having been driven over by a car. These publicity stunts created huge public interest for the Kirk Precision bike, which was exhibited at the New York and the Cologne cycle shows in 1986. Kirk subsequently received the financial backing he needed to go into large-scale production.

However, the frames suffered from bonding problems and many were returned after gear bosses came loose and threaded bottom bracket liners came detached. In addition to these issues, the frames flexed more than equivalent steel or aluminium frames and did not weigh any less, reducing the attractiveness of the design. The power company Norsk Hydro took a stake in the manufacturing company, injecting more money into development. But despite showing such early potential, Kirk's magnesium design – the most radical that had been seen in the bike industry for 25 years – never got past the initial teething problems to become mainstream.

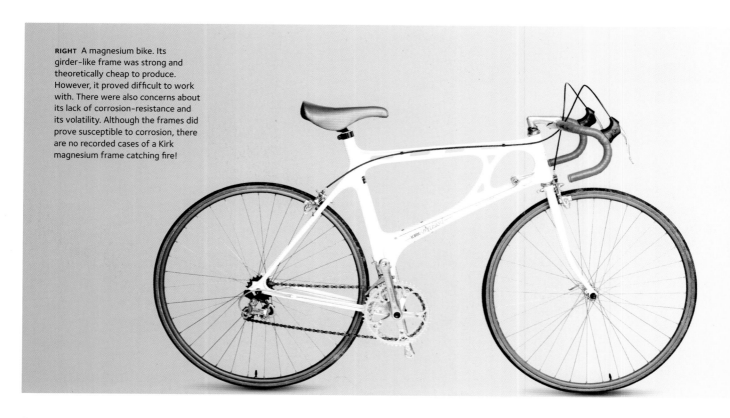

RIGHT A magnesium bike. Its girder-like frame was strong and theoretically cheap to produce. However, it proved difficult to work with. There were also concerns about its lack of corrosion-resistance and its volatility. Although the frames did prove susceptible to corrosion, there are no recorded cases of a Kirk magnesium frame catching fire!

Perhaps one of the wildest concepts was dreamed up by Ford car designer Frank Kirk, who had experience in magnesium casting. Kirk believed that because magnesium was abundant and could even be produced by electrolysis on sea water, it might be a good material for bike frames.

The beryllium bike >

Two-thirds more dense than aluminium, beryllium is an incredibly strong and stiff metal. In theory these factors should make it the ideal material for bicycle frames but two factors prevented it from becoming common in the cycling industry: cost and toxicity. Beryllium is a very rare element, making it approximately 200 times more expensive than aluminium. A beryllium frame (beryllium tubes bonded into aluminium lugs) was developed by the American Bicycle Manufacturing Company in 1993. They advertised the frame for sale for a staggering US$26,000. In addition, beryllium dust is toxic to about 2% of the population and can cause death so manufacturing was extremely problematic. Unsurprisingly, it didn't catch on.

Steel comeback >

When the bicycle first went into mass production in the late 19th century, steel was the material of choice. Since then, it has been in and out of vogue. However, every time steel seems to have been left behind, a method has been discovered to squeeze more performance from this ever-reliable material.

Since 2010, steel has been enjoying a resurgence in popularity among high-end bike manufacturers. The British steel tube manufacturer Reynold has a new steel tube, branded 953, which is made using a sophisticated cooling process that changes the crystalline structure of the metal as it hardens from a molten state. The result is a new alloy that is purportedly able to challenge the performance properties of aluminium, titanium and even carbon fibre (see pages 118–19 and pages 182–4).

LEFT/ABOVE In 2012 the bike manufacturer Genesis began to produce steel bikes in quantity. Their Volare range contains a model made from Reynolds 953 stainless-steel tubing.

Aerodynamic Riding >

There are a raft of complex rules and explanatory diagrams dictating what can and cannot be used in international competition and they are constantly being amended, added to and changed. In fact, through the first two decades of the 21st century the UCI regulations became so complex they were accompanied by a 'guide to implementation' – in effect, a set of rules on how to interpret the rules! Nevertheless, there is room for variety within the strict doctrine governing bikes and rider positions.

Low-down arms >

Apart from steering, cyclists use their arms only to accelerate and climb out of the saddle. Such activities constitute only a very small fraction of racing. For the rest of the time, the arms contribute very little other than to support the weight of the upper body. At speed, the arms – effectively vertical cylinders – create significant drag, so removing them or changing their position to horizontal (fist on, to the wind) significantly reduces drag and so the energy needed to go forward at a given speed. Graeme Obree achieved this first by tucking his arms up to his chest (see pages 202–3) and later by making his arms almost completely horizontal, with his Superman position (see pages 204–5). Both of these positions were effective but are now no longer competition legal.

The German Jan Ullrich, who won the Tour de France in 1997, used triathlon bars that lowered his torso; his weight rested on his elbows, and his forearms were horizontal. In addition, his arms were close together, in line with his thighs, effectively punching a hole in the air for his legs to pass through. The narrow elbow position also served to round off his shoulders, smoothing the flow over his back.

Ullrich's aerodynamic pointed helmet sat neatly on his curved back, filling in the low-pressure area behind his head and thereby transitioning the air across his shoulders.

Flat back >

Like Ullrich, the former world and 2008 Olympic time-trial champion Fabian Cancellara of Switzerland would sit on the very nose of his saddle, allowing his pelvis to rotate and his back to flatten. His arms were even closer together than Ullrich's, virtually touching one another, which gave him a very compact, rounded form. The air was forced to treat him and

LEFT The best time triallists might use very similar equipment, but they have subtly different riding positions. Each position is the rider's best attempt at being as aerodynamic as possible within the dimensions and limitations of their own bodies. There are also different schools of thought regarding head, hand and arm position. For example, the 1997 Tour de France winner Jan Ullrich, shown here, keeps his hands low, and lowers his head and torso towards them.

his bicycle as one shape rather than a collection of tubes. During an analysis of his time-trial position in 2009, Dr Andy Pruitt of the Boulder Center for Sports Medicine found that Cancellara was extremely flexible, which helps him maintain a close-elbow flat-back riding position.

Aerodynamics v power balance >

Although aerodynamics play a significant role in time trialling, the power that is put into the pedals is equally important. Consequently, a position not only needs to be aerodynamic but must also be one in which the athlete can produce power effectively. The optimal position is a compromise between one of aerodynamism and one of mechanical efficiency.

It is reported that Cancellara, one of the best time trialists in the world, changes his time-trialling position slightly depending on the course and duration of the event. For short time trials, such as prologues, he lowers his handlebars. For longer events, he prioritizes aerodynamics.

Narrow Wiggo >

Sir Bradley Wiggins is a very tall rider but also exceptionally slim. His arms and legs, although long, are at least a third smaller in diameter than those of his time-trial opponents. His track background with the Great Britain team as well as his time in a wind tunnel have given Wiggins a good understanding of the importance of aerodynamics. Consequently, a considerable amount of time has been devoted to training in a low tucked position similar to that of Cancellara. Perhaps the only small difference between them – apart from their size and body shape – is that Wiggins's head is brought low, towards his hands and in line with his torso. If seen front-on in silhouette, it would be hard to notice his head at all.

ABOVE Former world and Olympic time-trial champion cyclist Fabian Cancellara rides with his hands very close together, so that the air flows around them and passes between his upper arms and legs.

RIGHT Cancellara's successor as world and Olympic time-trial champion, Sir Bradley Wiggins, brings his arms up closer to his head, while keeping his elbows tucked in quite close, so that his arms and hands almost present one block to the airflow, rather than two.

Whatever kind of cycling interests you, there's a bike that's been adapted for doing it. Bikes for play, bikes to race on, bikes to explore with and bikes to use at work – inventors, designers and manufacturers are thinking about and developing them all. In doing so, they are refining what is already one of the most beautiful and efficient machines on the planet.

The Law-Makers >

Bikes you buy in the shops are governed by two sets of regulations: those that define bikes for racing and those governing safety. The world governing body for cycling, the Union Cyclist International (UCI), is responsible for the former set, while countries or groups of countries such as the European Union (EU) define the safety standards for cycling and ensure that manufacturers comply with them.

Manufacturers >

The EU safety rules are broad-ranging and put the onus squarely on each manufacturer to produce consistently safe bikes. There is a set of agreed standards that define what 'safe' is as well as the stringent set of tests each bike must pass before being deemed ready for sale.

In addition to passing these tests, manufacturers must supply all new bikes in the EU with wheel and pedal reflectors and a bell. The European Union rules also specify that in a county that drives on the right side of the road the rider's left hand must operate the front brake and their right hand the rear. Where they drive on the left, the opposite applies.

As manufacturers often sell bikes globally, they need to ensure that their products comply with other global standards, too. For the US market, bikes must have small flanges, often referred to as 'lawyer lips' (see picture opposite), which prevent the front wheel from coming out even if bolts or quick-release axles are undone.

In October 1973, bicycle advocate John Forester took on the American Food and Drug Administration – later the Consumer Product Safety Commission (CPSC) – over what he argued were ill-conceived cycling safety regulations regarding the positioning of light reflectors on bikes and, in particular, the stipulation that a white reflector be used instead of a front light. He sued the CPSC, which prompted a judicial order to review the regulations. The resulting reforms made them safer and more practical to enforce. Forester also rolled out a programme of cyclist education in America.

RIGHT/OPPOSITE LEFT The goods on sale in modern bike shops have to meet stringent safety standards. These have been developed over a number of years, not just by the authorities dictating the rules but with input from bicycle advocates, cyclists and cycling bodies.

John Forester – Bicycle advocate >

Much of the law governing bike manufacture and sales is derived from the actions of or lobbying by bicycle advocates. One of the most famous of these was John Forester, the eldest son of novelist C S Forester.

Born in Dulwich, south London, and a passionate cyclist from childhood, Forester moved to Berkeley, California in 1940. In 1973, he became a full-time cycling advocate, writing for various publications and arguing vigorously for segregated bike lanes in Palo Alto, California.

BELOW As the name suggests, the term 'lawyer lips' came from US lawsuits in which bike owners sued the bike manufacturers when their front wheels came out of the frame when the bike was in use.

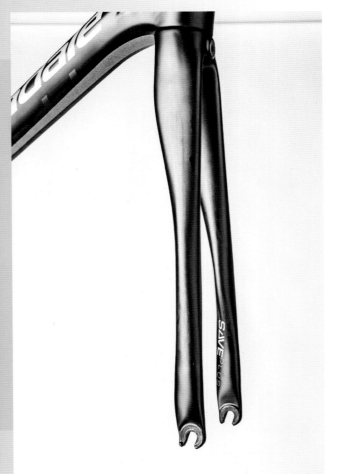

Bike Design Moves On >

Bicycle racing drives the development of ever lighter, more efficient machines. Racing bike designers, constantly searching for speed, introduced aluminium frames to save weight, index gears to improve efficiency and clip pedals for power transfer. These racing-derived advances are now prevalent on all types of bicycle. Although competition cycling will always be a big motivation for bike designers, other factors such as sustainability, environmental impact and ease of use are now influencing designers' thinking.

Racing bikes >

Today perhaps the single biggest area of focus in racing bike design is aerodynamics. In the early 1980s, when aerodynamics became a focus for bike designers, their attention was almost exclusively on time-trial bikes. It was decades before they realized that the shapes and methods they were developing for racing against the clock would also bring a performance advantage for riders in bunch races. With the arrival of carbon fibre, a material that could be easily shaped without increasing weight or losing stiffness, aerodynamics became a major focus of road bike design too.

Despite the industry's ability to produce bikes as light as 5kg that meet all current safety standards, the UCI's imposed minimum weight limit for racing bikes has remained at 6.8kg (15lb) since 2000. This enforced parameter reduced manufacturers' motivation to invest R&D funds in making lighter equipment; however this now looks set to change. At the time of writing the UCI is coming

RIGHT/BELOW One recent trend in road-race bike design is to improve their aerodynamics. This bike has flattened inside edges to its chain and seat stays [1], aerodynamically shaped main tubes [2] and a head tube that channels airflow optimally over the bike [3].

under pressure to remove the blanket limit and replace it with a safety standard. Just the rumour of this regulation's demise has rekindled the interest of manufacturers in making ever lighter bikes. Already new featherweight machines are starting to appear.

Mountain bikes >

Mountain bike designers, unconstrained by weight limits, fight a never-ending battle to shed grams. These all-terrain machines can be split into two broad categories: full-suspension and hardtail. Full-suspension bikes are heavier but can handle tough rocky ground, while hardtail bikes, with only front suspension, are used for fast racing over stony trails and tracks. Under each banner are several sub-categories. Trail bikes may only have front fork suspension but tend to have longer travel – around 140mm (5⅔in) – and are married to frames with slightly more laid-back angles, making them suitable for long days on tough trails. Cross-country racing bikes have short-travel suspension – around 80–100mm (3½–4in) – and harsher angles for faster handling. These characteristics are well-suited to race situations where reaction time and speed are paramount but tiring for all day excursions.

Perhaps the biggest change in the mountain bike world over the past few years is the introduction of the '29er' as opposed to the 26-in tyre. Large wheels certainly smooth out the trails and roll well over rough ground but they bring increased weight and slower handling. The debate still rages over which wheel is better. A halfway option, the 27.5', is gaining rapidly in popularity.

Utility bikes >

For many decades, a bike was a bike was a bike. Now designers have started to experiment with everyday machines, making them fashion accessories as much as practical modes of transport.

Many have gone for the minimalist approach, stripping away gears and even replacing the chain with a belt drive to reduce maintenance. Folding bikes have also exploded in popularity with brands like Brompton, previously the preserve of middle-aged gentlemen, now the must-have fashion accessory in London. All of these variations are a result of the current bicycle-for-transport renaissance.

Although the focus of these machines is utility, it does not mean they are unsophisticated. Many use exotic materials such as carbon fibre and titanium in their construction, while others have built-in electric drives and lighting systems.

From Drawing Board to Production >

My own company has a particular design philosophy, a holistic approach that combines the latest techniques and thinking with real-world experience. Each promising bike design, often starting as a simple sketch on a scrap of paper, is developed into a sophisticated computer model. These ideas are then tested, tweaked and optimized on a computer. Designing in this way is a technique derived from Formula 1 research and enables us to test hundreds of load paths, aerodynamic features and different carbon-fibre lay-ups before a single working prototype is made.

Computer-aided design (CAD) >
A specialist team converts the rough sketches into 3-D models, which allow my colleagues to see more clearly whether particular elements of the design will work. From this point, changes can be made digitally. As the airflow over a bike changes dramatically with the presence of a rider, laser-scanning techniques are used to capture perfect digital replicas of the bodies of some test athletes to feed into the computer modelling process.

Computational fluid dynamics (CFD) >
Once a CAD 'package' of frame, components and rider is assembled, they are tested using CFD virtual aerodynamic testing. CFD models air molecules bumping into each other and into the objects being tested, and is hugely complex, requiring enormous amounts of computer processing power. Once the tests are run – which can take several days of computing time – the design team reviews the models in visual form and identifies any areas for improvement. Many variations are tested this way until the team is happy to move on to the next stage.

Finite element analysis (FEA) >
Once a shape has been defined, virtual loads are applied to the bike in the same way that it would experience in the real world,

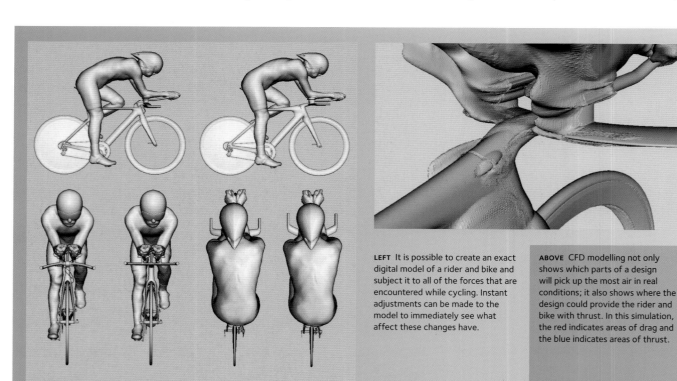

LEFT It is possible to create an exact digital model of a rider and bike and subject it to all of the forces that are encountered while cycling. Instant adjustments can be made to the model to immediately see what affect these changes have.

ABOVE CFD modelling not only shows which parts of a design will pick up the most air in real conditions; it also shows where the design could provide the rider and bike with thrust. In this simulation, the red indicates areas of drag and the blue indicates areas of thrust.

and the load paths are noted. With stress areas identified, the design team inputs the formula representing different material types, quantities and orientations until the required strength and stiffness are achieved. If required, the bike's shape can also be modified to distribute loads.

Wind-tunnel testing >

Having passed all the virtual tests, a prototype (or prototypes if the process reveals more than one viable option) is wind-tunnel tested with the same athlete used in the computer modelling. In this way, the wind-tunnel results can be compared to CFD tests, to check whether the computer modelling techniques are providing useful data. This is a self-validating process. If the wind-tunnel results match those from the computer models and the product is what has been hoped for, production prototypes are created. These are then field tested for a number of months before the bike is signed off.

Ignorance and expertise >

Einstein once said: 'Never ask an expert to innovate, they know what you cannot do.' My experience leading the Great Britain research and development team for ten years taught me that a top design team has a mix of expertise and complete ignorance. The latter is essential if you are to find the right questions. The required mix for a frame research project might be: a former pro bike rider, who understands handling characteristics and practical essentials; a composites specialist, who can recommend the best materials for a given application; and an aerodynamicist, who can advise on airflow matters.

The best results come when all of these individuals ask each other questions in complete ignorance. The result is cross-pollination of thinking, usually resulting in new and innovative solutions. Computers, CAD, FEA and CFD do not come up with ideas – they test them. Thankfully, there is still plenty of room for imagination and creativity in the design process.

RIGHT CFD modelling has replaced a lot of the design work that was formerly done in wind tunnels. However, many bike manufacturers still use wind tunnels to check computer-generated results. Wind tunnels are also used by individual riders who are trying to make their riding position more aerodynamic.

Rise of East Asia >

It is not unusual to see riders from different teams sponsored by three different bike brands stand on the podium of a major race, with all of their bikes having been produced within a few hundred kilometres of each other.

The majority of bicycle factories are now located in China, Taiwan and other East Asian countries and are known as Original Equipment Manufacturers (OEM). Some of these OEM's are absolutely huge but there are also many small carbon-fibre specialists accepting orders and sometimes working exclusively with only one or two specific bicycle manufacturers. Any cycling brand name can take its design to one of these factories and the factory will produce exactly what that particular brand wants – whether it be a set of tubes, a raw unpainted frame, a painted boxed frame or a ready-to-sell bike.

China bikes >

China is the world's biggest manufacturer of bicycles. It was therefore inevitable that when carbon fibre became fashionable (see pages 184–5) it would be among the first to adopt and master the art of mass-producing bikes made from this wonder material. By the 1990s, the majority of the world's carbon frames were being produced by just a handful of factories in China and neighbouring Taiwan. A decade later, many new factories specializing in carbon-fibre components for the cycle industry had sprung up to meet the demand of the bike industry. The biggest of these was Taiwan's Giant Bicycle Corporation, which, as well as making its own brand products, has manufactured bikes designed by a number of other big brands, including the iconic Italian bike producer Colnago. Many of the famous brands started their East Asia production in Taiwan but later moved to China to take advantage of lower manufacturing costs.

RIGHT/OPPOSITE East Asia is now the biggest centre of bike production, with factories manufacturing bikes to specifications provided by some of the biggest names in cycling. The factories work to the highest standards, producing raw or finished frames and complete bikes – whatever the customer wants. Where necessary, the large factories also draw on the expertise of small specialist carbon-fibre fabricators.

Worried by the ever-increasing number of bikes that are coming out of China at prices that European factories were unable to match, the European Union (EU) introduced the 'anti dumping' tax on products that had been manufactured in China. As a consequence of this measure, other countries, such as Vietnam and Thailand, who have a lower duty status, have become an attractive alternative manufacturing base for many global brands.

European alternative >

Not all famous bikes are made in East Asia, though. The carbon-fibre pioneer Look, for example, still produces many of its bikes in France, and the Bicycle Manufacturing Company (BMC) has invested very heavily in a Swiss factory where it uses the latest production methods to produce a few high-end models.

BMC >

BMC's Impec model is built in its Swiss factory utilizing robots for many of the production steps. BMC even makes some of its own tubes, weaving the carbon thread from scratch with machinery that was previously used to grade steel wire. Making tubes in-house allows the company to control the shape and make-up of each tube, customizing each for its intended use.

The frame tubes are then bonded together using high-tech injection moulded lugs. The method is not unlike the way in which the very first generally available aluminium and carbon-fibre frames were produced.

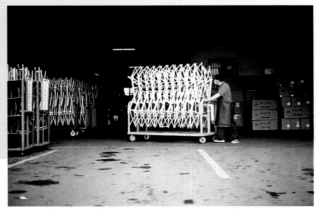

The Sportive >

For many years, serious cyclists were classified into tourers or racers and between the two was a huge social gap. In the 1980s, a new activity was created in Italy that bridged the gap between the two genres: the Gran Fondo.

Loosely translated, Gran Fondo means 'great foundation', 'great distance' or 'great endurance'. It can involve a full-on race, or just a bike ride. It is both a sporting event and a cultural experience and had been part of Italy's cycling culture for more than a decade – before the rest of the world finally caught on to the wonderful concept and re-branded the event as a cyclosportive or, simply, sportive, with events like the Étape du Tour (where riders did a stage of the Tour de France) which started in 1993.

The reason for the growing popularity is that each participant can make of the event what they want. Participants can treat it as a full-on race or just a day out with friends, stopping at every feeding station for a chat, while cycling over a fixed course. Participant's bikes are usually fitted with a transponder, which gives everyone a finishing time at the end of the day. Sportive routes are usually run on testing, often spectacular routes, and offer riders a choices of two or three distances.

Opportunity >

The cycle industry spotted the opportunity that sportives represented, and companies set about designing products specifically for the event. Many who enjoy this type of challenge like the idea of using the same bikes as the pros, but usually without a matching physique, the need for a low aerodynamic position or a desire for the harsh ride of a full-on racing bike.

A sportive-specific bike uses much of the technology of the racing machine but adds elements to make it suitable for long-distance riding by the masses. Frame weight – and costs – can be on a par with those ridden by the pros but angles have been slackened, head tubes lengthened – to give a higher front end riding position – and rear seat stays slimmed down, to enable road shock to be absorbed. In addition, most sportive bikes use compact chain-sets and larger sprockets, to give much lower gear ratios for slower climbing.

Another growing fashion with this kind of bike – currently not allowed in races sanctioned by the UCI – is the disc brake (see pages 144–5), which gives better all-weather stopping power. As pro bikes are restricted to a minimum weight limit of 6.8kg (15lb), many sportive models are far lighter than those used in the Tour de France.

RIGHT Cyclosportive bikes have longer head tubes than race bikes, so their riders have the option of sitting more upright than they could on an equivalent race bike. It takes racers many years to condition their bodies to crouch low for long periods of time, so for many cyclists sitting more upright is more comfortable over long distances. This bike still offers the option of a lower, more aerodynamic riding position if the rider simply holds the bottom of the dropped handlebars.

TOP LEFT Riders in the 2013 Étape du Tour. This cyclosportive event allows all-comers to ride the same route as stages of each year's Tour de France. The roads are closed to other traffic and the riders are supported by the organization with feeding stations along the route.

BOTTOM LEFT Cyclosportives are hugely popular in America. Here, entrants line up at the start of the 2012 Gran Fondo in New York.

Cross-pollination >

Technology flows in two directions and some sportive advances are now also used in dedicated racing bikes. Shock-absorbing features such as those used by Trek can often be seen in pro tour races like Paris–Roubaix or the Tour of Flanders, to reduce cumulative fatigue.

TOP RIGHT Fabian Cancellara of Switzerland leads this group at the Trek team's 2013 training camp in Altea, Spain. The bike he is riding, the Trek Domane, is as comfortable on smooth roads like this as on the cobbled roads of Northern Europe.

BOTTOM RIGHT Cobbled roads are the principal difficulty in two of cycling's most famous races, the Tour of Flanders and Paris–Roubaix. Here Cancellara leads Belgium's Sep Vanmarke over cobblestones in the 2014 Tour of Flanders.

The Fixie >

A bike seen in increasing numbers on city streets since 2000 and inspired by the minimalist bikes used by cycle messengers is the Fixie.

Messengers require simple, easy-to-maintain bikes that can nip through small gaps in the traffic. Unsurprisingly, they gravitated towards the simplest of all bikes – the fixed-gear track bike. Traditionally made of steel, track bikes are very light due to their lack of gears and having only one brake. The low weight combined with the direct-drive transmission gives excellent acceleration and road feedback.

Cycle messengers >
The tough world of the cycle messenger has received a lot of publicity since 2000, and their culture has had an impact on street fashion. One of those influences is the popularization of the track bike, which, with a few design tweaks, is ideal for navigating traffic-choked city streets.

Unlike pure track bikes used to race on velodromes, which have no brakes, Fixies have at least a front brake to be road-legal. This minimalist stopping power is augmented by using the fixed-chain aspect of the bike to put reverse pressure on the pedals. Some Fixies have a reversible rear wheel, with a fixed sprocket on one side and a free sprocket on the other. As the ability to brake via back-pedalling is lost when the free sprocket is used, these models need two brakes to be road-legal.

Flip-flop >
The reversible rear wheel – also known as a flip-flop wheel – is a new name for a type of hub that was used by the first road racers at the turn of the 20th century. It allows a choice of gears to be carried in the most rudimentary way. The two-sided hub also gives an option to freewheel, perhaps for descending or very technical city riding, where the inability to stop pedalling around corners may be hazardous.

Another way in which the Fixie differs from the track bike is its handlebars. On them, the brake is nearly always situated near the stem to giving an upright stance (for extra control when braking) and a narrow arm position (for riding between traffic). To aid riding on packed roads, many bikes also have the dropped part of the bars removed completely, leaving a super-narrow crossbar. Another popular configuration is to flip the cut-off bars over.

RIGHT A vintage Fixie. This classic steel bike is what many British cycle enthusiasts rode in the 1940s and '50s. Stripped down, as shown here, they could take part in time trials or hill climbs, even track races. Add mudguards and a saddlebag and it was perfect for long-distance and leisure rides, as well as commuting.

BELOW A stripped-down Fixie with upturned handlebars and platform pedals. This would be a great bike for scooting around town on short journeys. However, as it has no brakes, apart from the resistance that the rider puts against the fixed gear, it would not be legal to ride it everywhere.

ABOVE A dropped handlebar Fixie, with brakes this time. This bike has a flip-flop reversible wheel, giving the choice of fixed gears or a single freewheel. A bike with a single freewheel is referred to as a single-speed.

Fixie culture >

Fixie culture is worldwide. Members of this unofficial society meet to socialize, share ideas and to ride, celebrating the simplicity of their usually customized bikes.

The Mash-Up can be an impromptu or well-organized race through an urban environment in which people compete not for prizes but for the shear thrill of being different. Held each year in the Justin Herman Plaza, San Francisco, is the Red Bull Ride + Style event. It is one of the world's biggest Mash-Up gatherings, and it celebrates Fixie culture and urban art.

RIGHT A rider doing stunts on a fixie bike at the annual Red Bull Ride + Style event held in the Justin Herman Plaza, San Francisco. This is one of the biggest Mash-Up gatherings in the world.

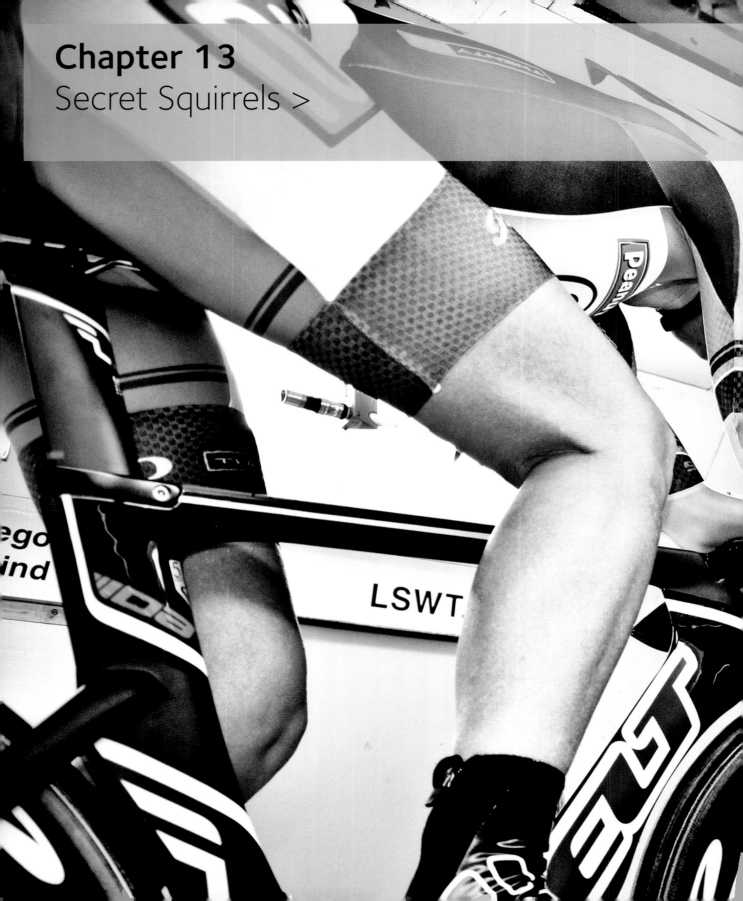

Most professional teams and many national cycling bodies use science and engineering tools and methods — like this wind tunnel — to enhance their competitive effort. Such work has become so important that teams who don't invest in research risk falling behind the competition before the race has even started.

The Minutiae of Speed >

As bikes improved and got faster, performance gains were harder to find. Disc wheels (see pages 164–7), triathlon bars (see pages 190–1) and aerodynamic helmets (see pages 196–7) had all helped achieve the big wins, but to move forward the cycling world had to change its philosophy.

The performance director of British Cycling from 2003 to 2014, Sir David Brailsford, coined the phrase 'the aggregation of marginal gains' to encapsulate the concept of effecting many small changes – rather than trying to find a single game-changing intervention – to make a meaningful improvement. It was a targeted approach, employed by the British Olympic team, that sought to speed up the evolutionary process slowly improving race bikes. The approach has turned up performance gains in surprising areas – those previously thought too insignificant to warrant attention.

Tyre turbulence >

A tyre can cause turbulence, which will add to overall drag on the bike as speed increases if it isn't carefully matched to the width and shape of the wheel rim it is mounted on. The ideal cross-section shape of a tyre and deep section rim would be a seamless teardrop where the transition from tyre to rim is smooth. If the tyre cross-section was bigger or smaller than the rim there would be a pronounced seam where they join, which would cause turbulence.

ABOVE Technicians at work in the Specialized wind tunnel prior to a test run on a road-race bike.

RIGHT Profiles of three deep-section wheel rims. Research has found that matching the rim profile to the size and even structure of the tyres used reduces turbulence.

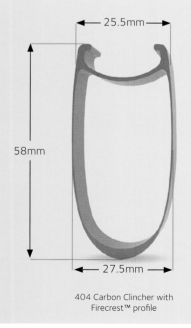

25.5mm

58mm

27.5mm

404 Carbon Clincher with Firecrest™ profile

404 aluminium/carbon clincher with hybrid toroidal profile

Traditional 'V' Shape

Looking deeper >

As aerodynamic performance is improved and drag drops, other factors such as rolling resistance become important. In the world of human-powered vehicles (HPV) racing (see pages 82–3), where vehicles now have a similar drag coefficient at 96km/h (60mph) as a car wing-mirror, tyre rolling resistance accounts for a huge percentage of energy use.

Rolling performance is made up of several factors including tyre width, tyre pressure, tread compound and the thickness of the tyre walls. Width dictates the shape of the tyre's contact patch with the ground and, along with pressure, governs how much the sidewall deforms. How fast a tyre deforms around varying surfaces and returns to its original shape is known as hysteresis. Wide tyres have a short, wide contact patch and return to shape very quickly, while narrow tyres with the same rider weight deform a lot more and absorb more energy. Although more aerodynamic, they tend to have a high rolling resistance.

ABOVE A bike set up in a wind tunnel ready for testing. The bicycle is mounted on a turntable, making it possible to assess both the effects of side-winds and head winds on its performance.

LEFT Sleek, modern, time-trial frames are the result of years of research, practical experience, wind-tunnel testing, CFD modelling and field trials. Here, the handlebars and seat post have been developed in conjunction with the frame, so that they work together optimally.

Designing a Time-Trial Bike >

Pete Jacobs of Australia won the 2012 Hawaii Ironman on a Boardman AiR/TT/9.8 bike. Despite this success, it could not be taken for granted that the bike used would offer the same relative advantage the following year. Being first at anything is a snapshot – a moment in time – so continued success can be assured only with a constant search for improvements.

Using techniques illustrated on pages 200–3, 204–5 and 224–5, the 2012 Hawaii success with Jacobs gave us at Boardman Bikes an excellent platform to build on and an excellent benchmark to measure progress against. Key to our success was the holistic approach to development; building the rider into all of the performance models and wind-tunnel testing.

This approach paid dividends when computer generated modelling techniques indicated that the handlebar/bike transition was a particular area for re-design focus.

Our first re-design of the handlebars occurred when computational fluid dynamics (CFD) modelled with bike-only showed significant gains. However, when the rider was added into the modelling, those gains disappeared. Further study of the CFD images indicated that the previously un-aerodynamic parts of the handlebar had effectively been punching a hole in the wind that the portion of the leg immediately behind then passed through. Making the handlebar more aerodynamic closed this hole, meaning the leg then had to create its own hole. We had reduced drag from one part of the bike only to create extra drag somewhere else. Being able to see this enabled us to tweak the design. Had we not brought the rider into the process, we might have invested a huge amount of resource for no performance gain.

Ending turbulence >
Another aspect of our development process is to use 'real world' conditions in our models. On the road, the wind is almost never head-on, so wind angles of up to 20 degrees were examined. The Boardman AiR/TTE was optimized to deal with a wide range of wind angles.

In the CFD models opposite it is easy to see just how much turbulence is created in the stem area and the savings that were made by carefully sculpting the bars and stem, blending them into the frame.

When marginal becomes maximal >
When all of the small design tweaks were added up and wind-tunnel tested using the same athlete who had appeared in the computer models for consistency, the Boardman AiR/TTE bike had 14–24 per cent lower drag than its predecessor, the AiR/TT, which had been built with the same equipment. When the rider was factored in and an average taken across wind angles, this equated to a performance gain – as opposed to just drag – of more than 7 per cent. Such a huge success shows just how much can be achieved by focusing on all of the marginal gains.

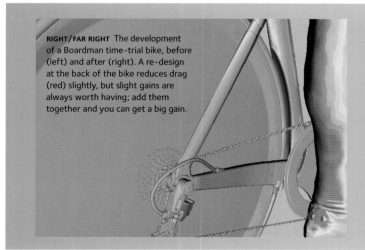

RIGHT/FAR RIGHT The development of a Boardman time-trial bike, before (left) and after (right). A re-design at the back of the bike reduces drag (red) slightly, but slight gains are always worth having; add them together and you can get a big gain.

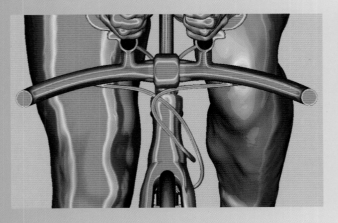

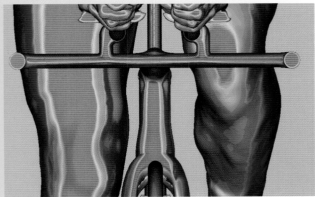

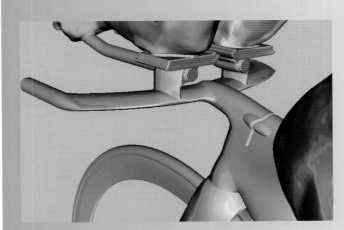

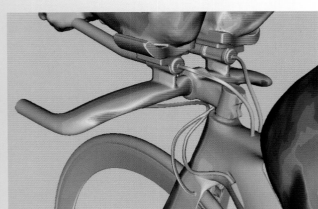

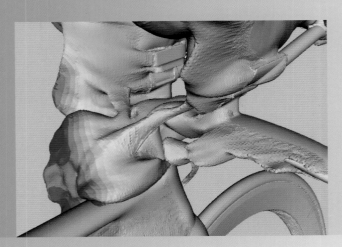

Bike Fitting >

A cyclist has three contact points with a bike: the handlebars, pedals and saddle. The position of these relative to each other can be adjusted to suit different physical characteristics. The aim is to maximize the rider's return – in speed – for the effort he or she puts in.

There are several broad brushstroke ways of setting up a bike to fit. Saddle height, for example, can be determined by the rider placing the heels of their bare feet on the pedals; the correct height is when their leg is absolutely straight at the bottom of the pedal revolution. Or you can measure an individual's inside leg to the floor in bare feet, then multiply that by 0.885 and use the resulting figure as the distance between the top of the saddle and centre of the bike's bottom bracket. Other elements, such as reach and fore/aft position of the saddle, would then be established. Using a plumb line, saddle position can be set so that the rider's forward knee is directly over the pedal spindle when the bike's crank is parallel to the floor.

Most general bike-fitting methods, like those above, do not take into account individual discrepancies such as differences in leg length. Neither do they factor in old injuries, people's lifestyle, body composition and so on. All of these factors affect a rider's positional preferences.

Specific fit >

A growing interest in understanding the mechanics of bike–rider interaction has led to a growth in bike-fit services throughout the world. Many bike shops now offer some form of fitting service for customers or offer it as a stand-alone service. Some bike manufacturers such as Specialized and Trek have created their own bike-fit systems for their dealers to use.

A bike-fit session often starts with the first points of contact – the feet and pedals. Then, once the feet are placed in the correct position on the pedals, the focus moves upwards in a process that London bike-fit specialists Cyclefit describe as 'chasing the kinetic chain'.

RIGHT A good bike fit should be personal, based on the client's needs, desires, shape and history. The measurements taken must be accurate, and the client should always be seen in action.

Bike-fit kit >

Three-dimensional motion capture technology – a technique pioneered in the film industry – was introduced to the bike-fitting work by American company Retül. Motion tracking points on the body are used to create a real-time stick figure video, which helps throw up issues such as limb misalignments, leg length discrepancies and even areas of muscle tightness.

The motion capture is not the only gadget that has found its way into bike fitting. Lasers are also used to check alignment, and some machines use memory foam to get an individual picture of pelvic bone structure so the correct saddle can be supplied.

Artisan frame-builders (see pages 118–19) use another approach to get the best frame fit. Rather than starting with a bike and changing components to alter position, they start with a fitting jig and adjust dimensions until the desired riding position is achieved. Just as a tailor takes individual measurements to create a perfectly fitting suit, the resulting measurements are then used to custom-build a bike frame.

RIGHT Bike fits today use a number of high-tech aids, ranging from lasers to motion capture. However, all the information acquired must be interpreted properly by the person carrying out the fit in order for its use to be effective.

Losing Weight >

In many race-critical situations – such as climbing a hill, attacking or sprinting – power-to-weight ratio (a measure of efficiency per kilogram) has a big influence on the eventual outcome.

Although far from being the only performance predictor, power-to-weight ratio is a reliable indicator of endurance form and is a number that reflects the maximum power a rider can sustain divided by their body weight, thereby giving a power per kilogram number. This ratio can be improved through specific training to increase aerobic efficiency and by losing body fat.

As the minimum bike weight is currently fixed by the UCI at 6.8kg (15lb), riders focus on themselves to shed every last gram of body fat. This often results in eating habits more often associated with supermodels.

Within that 6.8-kg (15-lb) weight limit there is still room for innovation. How weight is distributed around the bike is important. Weight saved on any revolving part is worth more than saving it on static element. The effect of low rotational

mass is so important that riders are prepared to use super-light carbon-fibre rims and sacrifice some braking efficiency in order to minimize weight around the extremities of their wheels.

How low can you go? >

The Lightbike, a creation of the German Gunter Mai in 2013, might not have an imaginative name, but at 2.96kg (6½lb) the moniker is certainly accurate. His bike was an amazing piece of engineering and, although only a one-off prototype, it served to illustrate just what could be achieved if restrictive regulations were removed.

Also in 2013 Scott produced the Addict SL, claiming it to be the world's first production sub-kilogram frame. They achieved this by mixing different grades of carbon fibre with expensive toughened resins and shaving small amounts of weight off all the ancillary parts. When built up with standard, commercially available parts, the bike was a fully functional road machine weighing 5.8kg (12.67lb), a full 1kg (2.2lb) under the UCI weight limit.

RIGHT The amazing Lightbike. Everything that could be done to reduce the weight of this bike has been done. Old fashioned, but very light, down-tube gear shifters and light brake levers take the place of integrated brake/gear levers, while the frame and forks are a one-off, super-light design. The remaining components are also the lightest of their kind.

ABOVE Although good road sprinters use light bikes, these riders generate too much power to ride machines that weigh as little as the Lightbike. Such bikes would flex under the intense effort made by sprinters, would be difficult to control and potentially dangerous. Sprinters need rigid bikes, which, by their nature, are much heavier.

LEFT The Scott Addict SL is well below the current UCI weight limit but Scott claim it is a fully functional road-race bike. Most authorities agree that the UCI weight limit is set too far on the side of caution.

Race Against Time >

The modern time-trial bike is a sleek machine, specifically designed for the job it needs to do. Bikes such as my own AiR/TTE are used by top male and female athletes in time trials as well as Ironman triathlons (see pages 210–11).

Adamo saddle designed specifically for riding in an aerodynamic tuck position.

Aero-section seat pos
a low frontal area an
flows smoothly aroun

Top tube is tapered a
flattened to improve

Flat-profiled s
tube for redu
frontal area a
improved airf

Rear wheel rim is deeper-sectioned than the front. It mimics some of the effects of a rear disc wheel but still offers the ease of handling of a spoked wheel.

Rear brake is located under the bike to shelter it from the airflow.

Electronic gear shifters are located at the end of handlebar extensions so the rider can stay in the aerodynamic tuck position when changing gear.

Base bars and stem designed specifically to smooth airflow over the rest of the bike.

Tri-bar extensions for an improved aero position.

Brake levers are placed at the ends of the base bars to ensure control and balance under braking.

Flattened narrow down tube improves aerodynamics.

Front brake located behind the fork crown and inside its carbon-fibre body to protect it from the airflow and prevent it causing extra drag.

Flat spokes are much more aerodynamic than round ones.

Deep section rims smooth airflow over the rest of the bike.

Racing Time in a Bowl >

Bikes such as this one are used to win Olympic and world championship medals in track events such as the team pursuit, individual pursuit and 1,000-m (1,093-yd) time trials.

Aerodynamic helmet covering the face, head and the space between the head and shoulders.

Custom-made aerodynamic integrated tri-bars, whose one-piece construction improves the aerodynamics of bike and rider.

The shape of this stem helps to smooth airflow over the bike.

Custom-made, carbon-fibre frame.

Front disc wheel. Riders in velodromes are unaffected by wind, so those competing in time-trial and pursuit races often opt for a front as well as a rear disc wheel.

Positioning the zip on the back of the skinsuit ensures that air flows smoothly across the front of the rider's torso. The zip is concealed to reduce disruption of the air flowing over the rider's back.

Power meter recorder stores data from each performance which is later analyzed by the rider's coach so the effect of any changes in fitness, equipment, rider position or event tactics can be quantified.

Track tubular bike tyres – the most expensive and lightest available.

Disc wheels are the optimum aerodynamic bike wheels. They are used in most top-level endurance events and some sprint events in indoor velodromes.

Drag Racers >

This is a track sprinter's bike. Of all bike racers – apart from motor-paced and human-powered vehicles (HPV) racing – sprinters achieve the highest speeds, so good aerodynamics are crucial in a sprint bike. As a result of the huge amounts of power and torque that these athletes produce, these bikes need to be stronger than any other type.

Deep dropped handlebars help riders to crouch low, thereby reducing their frontal area.

Deep-section aerodynamic rims improve airflow. This kind of wheel is easier to manoeuvre than a disc wheel; sprinters prefer to use at least a front spoked wheel to ensure an instant response from their bikes.

The absence of seams and creases in modern skinsuits improves rider aerodynamics.

Aerodynamically shaped seat post smooths airflow.

Teardrop seat tube reduces frontal area and improves airflow.

Flared seat stays improve airflow.

Ultra-light track tires pumped up to 120 PSI+. Fully inflated tires have a tiny contact patch with the wooden velodrome surface, which reduces their rolling resistance.

Riding the Tour >

The modern road-race bike is light, robust, easy to ride, responsive and comfortable. These are easy words to write but are difficult characteristics to build into a single design. Bikes similar to this have been used by the world's top athletes in that most famous of races, the Tour de France, since 2010. Despite their sophisticated design, their DNA is clear. Their basic form has changed little since the Tour de France was conceived more than 100 years ago.

Light road-race saddle reduces the bike's weight.

Aerodynamic seat post improves airflow.

Aerodynamically shaped frame

Electronic front and rear mechs for ease of use and improved accuracy of shifting.

Rear brake located under the bike to remove it from the airflow and reduce aerodynamic drag.

All control cables run inside the frame to remove them from the airflow and reduce aerodynamic drag.

Integrated brake levers and shifters ensure ease of shifting.

Di2 gear shift electronics. Cable connections are positioned under the handlebars to protect them.

Front brake located behind fork crown to remove it from the airflow and reduce drag.

Deep section rims improve airflow and ensure good bike handling.

Extra-long valves are used with deep-section rims for ease of tire inflation.

Flat spokes are much more aerodynamic than round ones.

Bikes have a huge contribution to make in the world. They are cheap, low-impact forms of personal transport, capable of carrying people and goods across vastly differing terrains. They are also a great way to get fit and explore, and provide us with a fascinating array of exciting sports.

Since the 1990s, cycle teams and individuals have increasingly looked outside of the sport for inspiration and knowledge to improve performance. It is not surprising that Formula 1 and aerospace companies, who specialize in aerodynamic understanding and composites construction, were the industries of choice for collaboration.

Perhaps the most influential tool of all was the wind tunnel. For the first time, athletes, coaches and manufacturers were able to get tangible feedback on the effect of their actions on performance. It was revolutionary. For manufacturers, the resulting information was used to guide equipment design, while athletes used the feedback to hone their riding positions.

Racing cyclists expend as much as 80 per cent of their effort overcoming air resistance, so it is hard to over-emphasize the importance of aerodynamics. By far the biggest part of the bike/rider package is the rider themselves (see pages 76 and 164). Consequently, a rider's position will have a huge effect on overall drag and performance (see pages 200 and 204).

Most of the world's top time trial and track riders have spent time in a wind tunnel to improve their position. Even those not occupied with riding against the clock – the ones most likely to reap the greatest reward from improved efficiency – have frequented these facilities. In 2010, road sprinters Mark Cavendish and Marcel Kittel of Germany both used tunnels to help improve their drag numbers.

Wind tunnels allow riders to see the impact of every change they make on drag. The drag data is usually translated for them,

LEFT Many cyclists, and not just professionals, take advantage of wind tunnels. This example, located in Brackley, Northamptonshire, was once used exclusively by a Formula 1 motor-racing team. Now any cyclist can pay for a session and use the data to fine-tune their riding position. Manufacturers use wind tunnels, too, when developing new models or refining existing ones, but many find Computational Fluid Dynamics (CFD) easier and quicker to work with.

OPPOSITE Turbo trainers are not just for training indoors during bad weather. Many models provide performance information that cyclists can use to quantify their training and help plan future sessions. Since there are no interruptions, such as slowing for corners or waiting for traffic, turbo training sessions can be very precise.

from coefficients and newtons of force, into time saved in their event. This feedback, often delivered in real time, drives their decision-making on everything from small changes in position to wholesale revisions to equipment.

Although highly informative, these facilities have drawbacks. They are typically very expensive to use, often difficult to access and, most importantly, the feedback they provide only tells the user what happened, not why. It can take a lot of time and effort – sometimes years – to unpick the cause of a given drag change. Because of this, many top manufacturers began using other tools used by the motor and aircraft industry, methods that allowed them to build and test virtual products on a computer, massively increasing efficiency, as well as saving time and reducing costs. (see page 224).

Imaging in 3-D >

Computational fluid dynamics (CFD) is a process by which the motion of fluids is mathematically modelled. It was first used in the 1930s as a way to test the aerodynamics of an object (even earlier versions of fluid modelling were used – unsuccessfully – to model weather). Since then, CFD has been transformed into a highly sophisticated science and is used extensively by both the aircraft and automotive industry to drive design. Unsurprisingly, CFD made its way into the world of Formula 1, and from there it has been picked up by bike designers.

Unlike in a wind tunnel, where the outcome is a number, CFD allows manufacturers to 'see' the airflow around the object they are testing. From there, they can quickly adjust a design and run the model again. Although not always accurate when applied to the relatively slow-speed world of cycling, these virtual images are highly instructive and often stimulate new ideas.

Virtual cycling >

For a host of reasons, from injury recovery to bad weather, cyclists have always needed an alternative to riding outside. The most basic apparatus to facilitate this, invented towards the end of the 19th century and still in use today, is rollers. These comprise a collection of three cylinders linked by a belt. The rider balances on these and their pedalling action allows them to steer and balance.

Turbo trainers are a more recent development. Most commonly, these are fixed to a bike via the rear-wheel axle and a strong spring holds a single roller against the rear wheel. Resistance – simulating the air resistance that a cyclist would normally encounter – is created via a small fan (hence the term turbo) while the inertia is simulated by a weighted flywheel. More sophisticated turbo trainers have magnetic resistance or a combination of air, oil and/or magnets.

Many modern turbo trainers have a digital display so riders can relate their efforts to the speed they would be travelling at if on the road. Some trainers give riders an indication of power output, allowing them to gauge workload and compare to previous training sessions. There are even trainers that use computer technology to generate a virtual world to ride through, displayed on a monitor or TV screen. With these, it is possible to programme race routes and create virtual competitors, with resistance automatically changed to simulate everything from gradients to downhill sections.

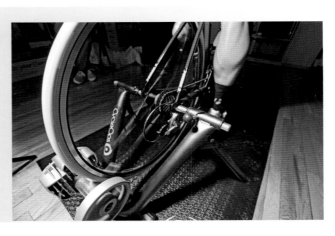

Going Electric >

In 1990, Japanese manufacturer Suntour began to experiment with electric gears but it was the French company Mavic who made the first serious attempt to create a mainstream system. Their ZMS (Zap Mavic System) was launched in 1992. It was ahead of its time, perhaps too much so, as even serious cyclists were suspicious of the technology. Operating only on the rear mechanism, it worked well and I used it on my time-trial bike for several years. Battery life was an issue and early production sets suffered from intermittent failures. The system failed to capture the public's imagination as anything more than a curio.

More work was needed to create a reliable easy-to-maintain-and-use, fully electronic system. In 1994, the German company Sachs brought out their version. Sadly, their Speedtronic gears were even more unreliable than Mavic's ZMS.

Determined to master the technology, Mavic worked on their system for several years and eventually launched Mectronic gears in 1999. Now completely wireless, it was even more sophisticated than the original. But with the complexity came more failures. Gear changes were slow and sometimes just never happened. Even the sponsored pros – of which I was one – did not want to use the system in competition and if they were disenchanted, then the public definitely didn't want take the risk. All through the first decade of the 21st century, Shimano and Campagnolo were experimenting with electronic shifting. Cycling geeks waited to see who would be the first of the big two to master the technology and win the bicycle equivalent of the space race. Shimano was first into orbit when they released their Di2 system in 2009.

Adaptable system >

Di2 shifters are solid-state switches located in the brake levers (and later at the end of the tri-bar extensions). The switches send signals by wire to a battery pack placed near the bottom bracket. The rechargeable battery pack supplies power to the derailleur motors, which move the derailleurs using a worm gear. Shimano claims that its battery pack can last up to 1,000km (620 miles) between charges. The system has an LED light to warn when it needs recharging.

Unlike with Mavic's Mectronic system, the rear derailleur shift times were on a par with mechanical systems, while the front derailleur shifts were almost 30 per cent faster than Shimano's mechanical counterpart. Traditional front shift systems needed some mechanical sympathy from their users, requiring them to ease their pedalling while the shift was made. This was particularly the case when there was a large difference in chain-ring size. The electronic system did not have such problems. Shifting electronically was very quick, sure and, once set up, there was no need for readjustment due to cable stretch. Faster, easy to maintain, more reliable, precise – electric gears had come of age.

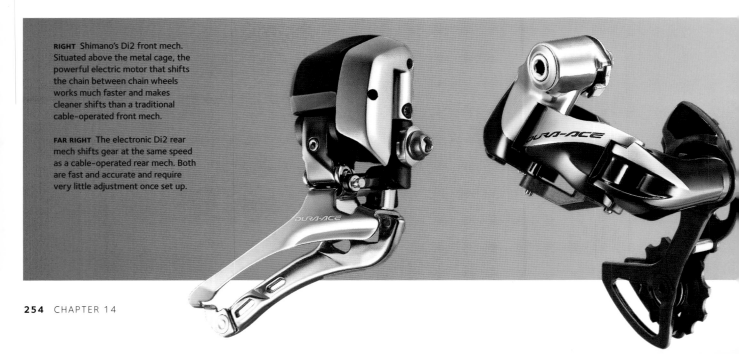

RIGHT Shimano's Di2 front mech. Situated above the metal cage, the powerful electric motor that shifts the chain between chain wheels works much faster and makes cleaner shifts than a traditional cable-operated front mech.

FAR RIGHT The electronic Di2 rear mech shifts gear at the same speed as a cable-operated rear mech. Both are fast and accurate and require very little adjustment once set up.

Campagnolo's answer >

At the end of 2011, Campagnolo finally released its own electronic shift system called EPS (Electronic Power Shift). The company had designed its electric system to allow riders to operate their gears exactly as they would their cable shift siblings.

EPS and Di2 have both advanced bicycle gear shifting and the competition between the two giants is making electronics more accessible to an ever-wider cycling market. It's not unreasonable to speculate that this is the start of a gear revolution and one day electronic gear shifts will be the standard for all bikes.

RIGHT/BELOW Campagnolo's EPS front mech, battery pack and combined brake lever/gear shifter unit. A lot of work went into the shifter to ensure that it shifted the chain at the same point in the arc of each lever as Campagnolo's cable systems.

ABOVE From the top: a Shimano Di2 battery charger; a lithium-ion battery that fits inside the bike's seat post; a lithium-ion battery charger; and a junction kit.

Charting your Course >

Not so long ago, cyclists who wanted to explore had to carry their maps in pockets and saddlebags, frequently stopping to track their progress. Alternatively, they clipped their maps to handlebar-mounted boards and chanced a glance down now and then. Those days are gone...

The Garmin Forerunner, the first GPS system that could be used by cyclists, came on the market in 2003. Today, handlebar- and even wrist-mounted GPS systems make it easy to explore new places by bike. They also provide a wealth of data for any cyclist in training. There are even apps that monitor information from a ride, automatically synchronize the data with a database and allow the rider to race against other people without them even being in the same location at the same time.

Routes can be downloaded into the GPS unit and followed through a digital display, either through direction indicators or by using a live map, just like a motor vehicle GPS system. It is possible to follow the courses of famous bike races in this way, and to share routes with friends.

GPS units from market leaders like Garmin also display and store useful data, such as current and average speed, ride time, distance, altitude and, in some cases, an estimate of power output. Some GPS units also monitor heart rate, while others work with power meters to display accurate power information as well.

RIGHT This Garmin GPS can be fitted to a handlebar mounting. The ring on the mounting goes around the handlebars and is bolted in place. The device is then easily attached to and detached from the mounting.

BELOW Three Garmin GPS devices showing the variety of readouts that are available. The device on the left is in map mode, so a rider can follow pre-determined routes while riding. The device in the middle is set up ready to ride. The device on the right shows the sort of information that is available during a ride; in this case current speed, heart rate, cadence and power.

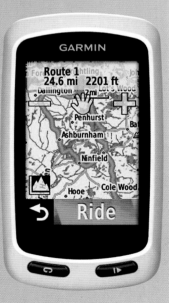

Virtual race >

Strava is an American company that in 2009 came up with a free app, which cyclists (and runners) can use to download their training rides and see how they compare with other users along stretches of road or trail (now widely referred to as Strava segments).

App users create the segments. The practice has proved so popular that it is as if Strava were dividing the world up segment by segment. Famous climbs like Alpe d'Huez and Mont Ventoux have been made into segments, and almost every hill a bike race has been up has been listed. Anything a user chooses can be declared a Strava segment, provided it is not deemed too dangerous.

Used sensibly, the Strava app is a great training tool. Not only does it encourage users to train to improve their times, but it also holds information on previous attempts, so over a period you can use it to track improvement.

BELOW A smaller Garmin GPS that can be worn on the wrist or fastened around a handlebar mounting. Many riders download rides from their Garmin or other GPS devices to the Strava app. This allows them to analyze their ride and see how fast they are compared to other riders on various stretches or 'segments' of the road.

Reviewing your ride >

Bike-mounted cameras are another device that has exploded in popularity since 2010. These can be mounted on handlebars or helmets. GoPro's Hero 4 camera weighs just 84g (2½oz), and gives TV broadcast-quality video as well as sharp, still pictures. Cameras that allow riders to look back at exciting rides and share the experience with friends enhance the fun of the outing. More seriously, if involved in an accident, they can provide invaluable evidence, too.

ABOVE Helmet- or bike-mounted cameras are small, light and robust. Modern units provide an exceptionally high-quality video record of a ride.

Electric Bikes >

When is a motorbike not a motorbike? When it's an electric bike. An electric bike is still pedalled using the normal drive train but gives the user the option to supplement their effort with power from an electric motor. These machines have to comply with often extensive regulations but, in the case of countries within European Union, they don't require road tax, insurance or a driving licence.

Ogden Bolton Jr was granted a patent for a battery-powered bicycle back in 1895. Then, in 1887, Hosea W Libbey of Boston, Massachusetts designed an electric bike that had a double electric motor in the bottom bracket and was powered by a large lead acid battery suspended between the frame tubes. Later designs stored energy from when the bike was ridden downhill or freewheeled to recharge the battery. The energy could then be used by riders if and when they needed it to augment their own pedal power. Other designs were based on electric motors that could be fitted to a range of bikes so they provided some pedalling assistance.

Multi-use bikes >

Electric bikes are very useful for commuters, older people and less experienced cyclists who want to extend their riding range. They are a great way to get fit because you have to pedal, but the motor helps on hills, allowing the electric cyclist to ride further.

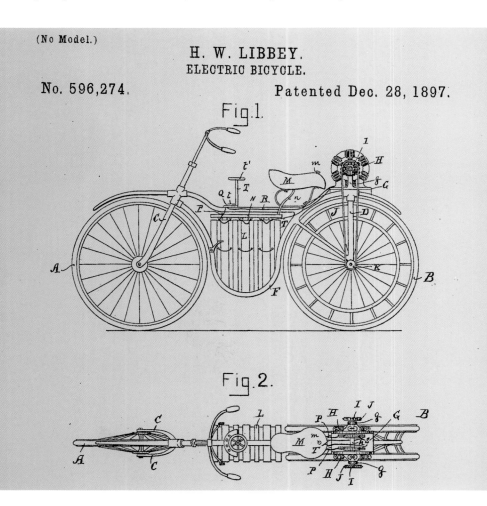

RIGHT Drawings of an early electric bike. Unlike its modern equivalent, this machine is powered completely by electricity, and there is no facility for the rider to pedal. A massive battery pack is attached within the bike's main frame tubes, filling the space between them, while the motor is mounted directly above the rear wheel.

(No Model.)

H. W. LIBBEY.
ELECTRIC BICYCLE.

No. 596,274.

Patented Dec. 28, 1897.

Fig.1.

Fig.2.

Electric bikes are very useful for commuters, older people and less experienced cyclists who want to extend their riding range. They are a great way to get fit because you have to pedal, but the motor helps on hills, allowing the electric cyclist to ride further.

Although an electric bike enables a cyclist to go faster, there is no danger of it getting out of control. To comply with European Union law, the motor has to cut out once the rider's speed exceeds 25km/h (15½mph). Electric bikes in Australia are regulated in a different way. Classed as power-assisted pedal cycles, the power output of the motor must not exceed 200 watts but speed is not controlled. In Canada an electric bike can have a motor of not more than 500 watts and be capable of no more than 32 kph (20mph) The speed limit is the same in the United States, although a power output of 750 watts is allowed. In China an electric bike is one that can be pedalled but is capable of no more than 20kph (12mph) when not being pedalled. China is the world's biggest producer of electric bikes. In 2007 electric bikes were estimated to represent between 10 and 20 per cent of all two-wheeled vehicles used on Chinese roads. Sixteen to 18 million units were sold nationwide in 2006 alone.

The Kalkhoff Pro Connect Xion X27, a German bike launched in 2012, is a great example of an electric bicycle. It has 27 gears and is designed for commuting, leisure cycling and more adventurous touring. Its electric power is delivered directly to the bike's cranks. The rider selects the amount of power assistance they need, which they can monitor on a handlebar-mounted display. The display also tells them how fast they are riding, how much charge there is in the battery and the bike's projected range at any given moment.

RIGHT/BELOW Modern electric bikes use electricity only to boost the rider's pedal power. Batteries and motors are quite small and unobtrusive, and the rider controls the amount of electrical power that is supplied. Some models also store usable energy that has been produced by the bike's motion.

Trickle-Down Effects >

A lot of features seen in modern road cars today started out as a performance-enhancing feature in the world of Formula 1. The same trickle-down of technology can be seen in cycling, where cutting-edge features seen in the Tour de France will often filter through to the bikes available in the shops.

Once aluminium was successfully introduced into racing and TIG-weld procedures (see page 182) became less costly, it became viable to mass-produce lightweight aluminium frames. Nowhere in the cycling industry is technology transfer more evident than when carbon fibre was introduced (see page 184). The performance gains of the wonder material were enormous: weight plummeted while strength increased but not without challenges. Carbon needed significant amounts of research and development as well as new and more expensive manufacturing methods. However, manufacturers knew that if a commercially viable production route could be found, the

potential return was high enough to warrant the required investment and risk. At first, the material was used only on top-end bikes. As production methods became cheaper and manufacturers more prevalent – particularly from 2008 onwards – carbon-fibre bikes came within the financial reach of even the occasional weekend rider.

After a short phase in the 1980s and 1990s of simply rolling the carbon fibre cloth into tubes, manufacturers realized that the way to get the most out of this material was to invest in moulds and make bike frames in one piece. Although the frames for the pros were of high-grade carbon fibre and expensive resins, their moulds could be re-used with less sophisticated materials to produce extremely cost-effective frames.

As a large part of the expense of carbon manufacturing is labour, it made sense to site factories in areas of the world where labour costs were low (see page 226). These factories now produce carbon-framed bikes that are comparable in price to aluminium ones.

RIGHT Although modern hybrid bikes make excellent vehicles for commuting and general riding, they benefit from the equipment and materials that were developed on racing bikes.

OPPOSITE Australian pro-racer Phil Anderson was the test pilot for Shimano's STI integrated brake and gear-shift lever units in 1989. STI was released the following year as part of Shimano's top-of-the-range Dura Ace group-set. All of Shimano's group-sets now have STI shifters.

Equipment >

Equipment has largely developed along the same tried and tested path as frames. New technology is developed for the racing elite, then, having gained publicity and generated demand, the volume sales justify trickling the technology down into more affordable products. This development and commercial strategy is clear to see in the story of integrated gear and brake shifters (see page 124).

Shimano's STI system was field-tested in the heat of pro race battles by the first Australian racer to wear the Tour de France yellow jersey, Phil Anderson. A Shimano mechanic, assigned to follow him for a year, used Anderson's comments to refine the design and make adjustments as needed. In 1990, the STI system was introduced on the Dura Ace group-set and over a period of years went on to appear across their entire equipment range.

Campagnolo went through a similar process with its Ergo-Power brake-shift units. If you look carefully at pictures of riders using the first version of Ergo levers in 1992, you will see a small elastic band holding the shift and brake levers together, to prevent them rattling. As a result of the pro's feedback, production models were fitted with a small rubber stop on the back of the brake lever.

ABOVE XX1 is a revolution in shifting for off-road bikes. Eleven different different gear ratios are available at the flick of a single shift – ratios that can cope with almost every terrain.

Mountain bikes >

Off-road bikes also used the elite ranks to test and refine product ultimately destined for the masses. Everything from shifters to suspension was trialled with the pros before going into more widespread use.

The pro riders have been used to do more than just test products. They have also helped influence trends. An example of this is SRAM's XX1, which was launched in 2012. In a sphere where three chain-rings were common place (see page 136), the XX1 gear system broke tradition by employing a single front chain-ring with a wide gear spread at the back, to accomplish the same task.

The front chain-ring lost some teeth to become a tiny 32 or 34. At the rear, the 11-sprocket chain-ring spanned a massive range from a minuscule 10-tooth bottom sprocket to an enormous 42-tooth dinner plate. All this was made possible by a very cleverly engineered rear mech and a custom pitch chain (a custom-made chain of slightly different dimensions to standard).

This approach was used for many years by cyclo-cross riders who wanted lighter bikes to carry and fewer places for mud to clog the system. In the mountain bike (MTB) world, however, it was introduced only when cyclists saw images of riders using the XX1 gear system in 2012 to win MTB Cross-Country races. This gave the consumer confidence in the new approach. Without that endorsement, it is doubtful whether cyclists would have got behind the concept. The X11 gear system now looks set to make its way down the range.

Still Serving >

Where do old bikes go? Some continue to go the way of all metal and meet their end in the scrapyard, but now, many do not. Noble old bikes are being pressed into service to improve and even save lives in developing countries.

Bikes are highly adaptable and have an obvious transport role to play over a wide range of terrain (see page 70). But the most effective machines will always be those specifically designed for their intended use.

Delft University of Technology in the Netherlands created two bikes based on experiences in developing countries. These were made to be able to last a long time with minimal maintenance and tolerate rough terrain. One model, based on an old transport bike popular in the Netherlands for 50 years, had a large cargo rack at the front. Since its first appearance in 1984, more than 100 models have been in daily use in several African countries. This bike, part of the university's African Bicycle Design Project, weighed 48kg (106lb) and was capable of carrying 75kg (165lb) at the front and 100kg (220lb) at the rear: a total capacity of 175kg.

The second model – the bush bike – was developed as the result of a study in Tanzania in 2009 and is only at the test stage. Its cargo racks are situated at the front and back of the bicycle and can carry a massive 150kg (330lb) at the rear and an additional 50kg (110lb) at the front.

Multiple uses >

In Malawi, bikes are used to pump water from wells and lakes, both to drink and to spread on the land using bike-powered pumps. There are also bike-mounted battery chargers. The charger bikes are ridden from village to village, and locals bring out any batteries they need replenishing. The rider places the bike on a portable stand (which raises the back wheel off the floor), plugs in the adapter and pedals away until each battery is fully charged.

RIGHT The White Mamba, which was designed as part of the African Bicycle Design Project by Joep Oberendorff. The front and rear carriers and long wheel base make it possible to carry heavy loads safely, even over relatively rough surfaces.

Bikes are highly adaptable and have an obvious transport role to play over a wide range of terrain. But the most effective machines will always be those specifically designed for their intended use.

Poverty relief >

Surveys done in African countries and Sri Lanka identified that bike ownership can increase household wealth by as much as 35 per cent. Bikes were also found to be an effective solution to road congestion in India.

To help reduce poverty, World Bicycle Relief – a not-for-profit organization based in Chicago – was founded; it distributes old bikes throughout the world to help people in countries less fortunate than our own to prosper. They are also used in disaster-recovery projects. By the end of 2010, the organization had redistributed 116,000 bicycles.

World Bicycle Relief also designs bicycles specifically for the environment in which they will be used. These are branded as Buffalo bicycles. The bikes are designed, tested and assembled in Africa and each has a coaster brake (see page 120) and weighs 22.6kg (50lb) including the rear rack and mudguards. Buffalo bikes can carry a cargo of 100kg (220lb).

True recycling >

Re-Cycle Bikes to Africa is another charity that refurbishes old bikes and redistributes them. They work specifically in Africa where people face enormous transport problems, and their lives are often limited by how far they can walk. Bikes cut down journey times and increase a person's range. This enables young people, who might otherwise not be able to get there, to attend school. Bicycles are particularly helpful for young women, who do much of a family's food transport. A bike reduces transport time, it leaves woman less vulnerable and it can carry loads, eliminating the health problems associated with carrying heavy loads in the traditional head-mounted way.

At the end of May 2014, Re-Cycle Bikes to Africa had shipped 54,317 bikes in 143 containers to Africa. You can find out more about their work on www.re-cycle.org, and about World Bicycle Relief on www.worldbicyclerelief.org.

ABOVE The Quebeka Buffalo Bicycle. Quebeka is an African project that gives bikes to children in exchange for their work to improve the local environment. The project receives worldwide publicity as a result of its joint sponsorship of the MTN-Quebeka pro-cycling team, making its Tour de France debut in 2015.

ABOVE No matter how unsuitable it might seem, a refurbished bike will get used by somebody in Africa, and it can make a big difference to that person and to their family's life.

A concept bike for the future. Just like concept cars, concept bikes are more about a designer imagining possibilities than a genuine idea for a production bike. However, some of the ideas that are dreamt up find their way into everyday designs.

Bicycle Utopia >

My personal passion for cycling, which might come as a surprise to many, is not with technology and racing but with bikes made for transport. The potential of this wonderful machine to solve many of the big problems our society faces today, from pollution and congestion to health and social well-being, is huge...if we let it.

Research has shown that two-thirds of Britons want to ride a bike more and would do so if they felt safe. Since the 1950s, when private car use became available to the masses, our streets have been designed around the needs of motor vehicles. For decades planners have, unwittingly or uncaringly, designed cycling and walking out of our infrastructure, discouraging people from travelling in ways that would most benefit ourselves and our communities (see page 104). Our reward for this short-sightedness is an environment in which children can no longer ride to schools, local parks or visit friends by bike. We now have 37,000 deaths a year in the UK directly related to obesity, and 15,000 deaths a year from pollution. In fact, the combined costs to the UK from inactivity is estimated to be just under 1 billion pounds...a week! Fewer than 2 per cent of children ride to school in Britain whereas a few hundred miles away in the Netherlands 50 per cent of children travel to school on a bike.

There is now a growing movement in the UK pushing for change, to get us to use the best practice examples of our European neighbours to change the way our streets and cities look and to prioritize people over traffic. I for one want my kids to be able to ride to school and to the park. I would like us all to be able to pedal to the train or bus station and to the shops. This will only happen if we make space for people to travel by bike.

The key to making this happen is two-fold: leadership from central government backing – and sometimes requiring – local authorities to change; and meaningful sustained levels of targeted funding to make walking and cycling the preferred means of transport for short journeys.

European foresight >

There are many good examples of how other countries have tackled this issue. In Denmark the authorities have provided bicycle superhighways – which have increased the speed, safety and comfort of bicycle commuting, linking them to other transport hubs to give people the option of using bikes as part of longer journeys. The first Danish bicycle superhighway, C99, opened in 2012, and runs between Vesterbro (central Copenhagen) and the suburb of Albertslund. The route is over 17km (10½ miles) long and has service stations with air pumps at regular intervals along

BELOW More space in cities is being given over to bikes, as many studies have shown they are the quickest way to travel around. This bike lane is denoted by broken white lines.

RIGHT A public bicycle pump next to a German cycle route. Simple facilities like this encourage people to cycle instead of drive. It's all about making bikes more convenient than cars.

In Denmark, the authorities have provided bicycle superhighways, which have increased the speed, safety and comfort of bicycle commuting, linking them to other transport hubs to give people an option to use bikes as part of even long journeys.

it. At junctions there are handholds and running boards so cyclists can wait without even having to put their feet on the ground!

Segregated cycle lanes in Karlsruhe, Germany are called Fahrradstrasse, meaning bicycle street. A network of similar segregated lanes run through other German cities, too, as they do in neighbouring Belgium and the Netherlands. At the end of the journey, secure bike parking facilities are provided to encourage people even more to choose this method of travel.

Link-up >

One of the reasons people sometimes leave the bike at home is the lack of joined-up options, so it was a welcome move for bicycle advocates when, in 2007, the European parliament passed a law stating that all international trains must have the capacity to carry some bicycles. It is a move that reflects the approach seen on a smaller scale in many parts of Europe.

The Rheinbahn transit company in Düsseldorf permits bicycle carriage on all its bus, tram and train services at any time of day. In France, the TGV high-speed trains will carry bikes if there is a bicycle icon indicated on the service when a seat is booked on-line. The bike is also booked at the time a seat is reserved.

Rest of the world >

When authorities in Japan started expanding bicycle parking at railway stations they noticed that if they made cycling easy and convenient, people would happily change their habits. In 1977, parking spaces in that country totalled 598,000 spaces; by 1987, this had risen four-fold to 2,382,000 – a number that includes no fewer than 516 multi-storey garages specifically for bikes.

All public transit buses in Chicago and its suburbs allow up to two bikes at all times. The same is true of Grand River Transit buses in the region of Waterloo, Ontario, Canada. In Victoria, British Columbia and in the Australian capital, Canberra, some buses have externally fitted bike carriers, enabling passengers to ride to a bus stop then hook their bike on to the carrier, travel in the bus and then use the bike to reach their final destination. The Canberra system was introduced in 2009 and expanded in 2010.

ABOVE Bicycle parking at Utsunomiya station in Japan's Tochigi prefecture.

This main-line station is located 75 miles (121km) north of Tokyo.

ABOVE Some bus companies around the world are fitting bike racks to their vehicles, so that passengers can take their bikes with them on the bus to use at their destination.

Saving the planet >

There are many more schemes around the world other than the ones mentioned here but if we are to make meaningful reductions to pollution and improvements in health, there needs to be even more. If they are allowed to be, bicycles have a huge role to play in our transport future. It is no exaggeration to say that the humble bicycle could make the world a better place to live.

Drive-Train Options >

Where the bicycle will go next and what the bike of the future will look like are subjects of continued speculation. One would think that, with such a simple principle, it would be hard to find improvements. However, if history teaches us anything, it is that as long as people have an imagination there will be scope to enhance this wonderful device.

The safety bicycle (see pages 24–5) was chain-driven. In fact the chain, around since the start of the industrial revolution, is such an efficient system it is still used on most bikes today to transfer leg power to the rear wheel. It might be adaptable and efficient but it does require regular maintenance and it can get messy. So is there a cleaner, lower maintenance alternative to the mighty chain?

Belt drive >
The Trek Soho used a carbon-infused rubber belt with teeth that fitted into indentations made in a special chain-set and sprocket.

The sprocket then drove the rear wheel through a hub gear. Although less efficient than the chain and requiring a more complex frame to fit new belts (unlike chains, belts are typically made in one piece), carbon-infused rubber belts still offer several advantages.

When the Soho was first released in 2009, Trek believed that belt-driven bikes would be more acceptable to the people using them because they were quiet, mechanically simple and required very little maintenance. Trek claimed that the belt drive resisted wear better than a chain and sprockets and, as it did not require lubrication, the possibility of the rider getting oil on their clothes was removed.

RIGHT A modern belt-driven bike. With its disc brakes and hub gears, this would make a great, low-maintenance commuter bike.

OPPOSITE A shaft-driven bike. Although ingenious, less prone to wear and requiring less maintenance, shaft and belt drives are unlikely to replace the roller chain in the foreseeable future.

Shaft drive >

Like belt drive, shaft-driven bikes have been around for a while, with a UK patent for a shaft-driven bike traceable back to 1891. However, because it was competing with the very efficient and well-understood chain drive, interest died.

The shaft – more properly known in engineering terms as a crankshaft – is turned by pedalling two opposing cranks. These move a bevel gear, which transfers the crank's axis of torque through 90 degrees. At the opposite end, the same thing happens: the crankshaft turns another bevel gear on the rear hub, changing the axis of torque back through 90 degrees to spin the rear wheel.

Shaft drive is low maintenance and simple to maintain. However, there is a lot of frictional loss in the system, making it a lot less efficient than chain drive; it can also make removing the rear wheel awkward. Like belt drive, shaft drive is cleaner and can be totally encased. This removes the possibility of

getting oil on clothing and things being caught in the drive when the cyclist is riding in ordinary clothes. These attributes would seem to make it ideal for utility cycling where practicality is valued more highly than efficiency.

Reign of the chain >

The chain is a hard act to follow and, certainly with current technology, it is hard to beat as a method of transferring leg power to the rear wheel. Belt and shaft drives have some advantages in certain arenas, such as commuting bikes (see pages 104–5), but if the chain does ever have a successor it is unlikely either of these mechanisms will get the role.

Rolling Along >

Re-inventing the wheel might be a tall order but it might be possible to improve it...

The Copenhagen wheel >

Using standard fittings, the Copenhagen wheel is effectively a self-contained electric power unit, able to turn any bike into a power-assisted hybrid. Inside the enormous rear hub are some sophisticated electronics and a regenerative brake. As well as slowing the bike when required, the brake harvests energy when the rider slows or freewheels. When sensors in the pedals detect an increase in power being applied by the rider, an electric motor kicks in and uses the stored power to supplement their effort. Via a phone app, the rider can change how the stored energy is used.

Going hub-less >

Doing away with spokes and hubs completely, the hub-less wheels are the most futuristic looking of all. Their rims are held in cradles that are part of the bike's frame, and drive from the pedals is delivered directly to the rear rim via concealed gears rather than a chain. Braking is achieved via a mechanism inside the cradle. John Villarreal's famous design of 2010 has dual tyres, with drive being applied to a toothed ring nestled between them.

The loopwheel >

Rather than use suspension systems built into the bike frame, the loopwheel incorporates the damping mechanism in its construction. Three leaf springs, each bent into an oval, replace the spokes between the hub and rim. Because the rear wheel carries more load and transfers energy from hub to road, the rear springs are less pliant.

In production since 2011, loopwheels work particularly well on small-wheeled bikes, which give a notoriously hard ride. As the loopwheel dampens road shock and vibration within its structure, narrower, hard tyres can be used, thereby reducing rolling resistance.

Pump tyres >

Invented by San Francisco's Benjamin Krempel in 2012, the PumpTire system consists of a tyre, an inner tube that clips into it, and an air valve. Air is drawn through the one-way valve, which protrudes from the rim like a regular valve stem. Instead of going directly into the inner tube, however, the air goes into one end of a small tube running along the centre of the tyre. As the tyre rolls against the ground, the tube compresses, forcing air out of its other end and into a second valve on the inner tube. The resulting absence of air in the tube creates a vacuum effect, drawing more air in through the first valve.

The valve is set to sense when the proper pressure has been reached, at which point it closes. As the pressure drops, because of seepage that occurs with all tubes over time, the air intake resumes, keeping the type at a constant pressure without any user input.

If the product works as planned, cyclists need never have to check or 'top up' their tyres again.

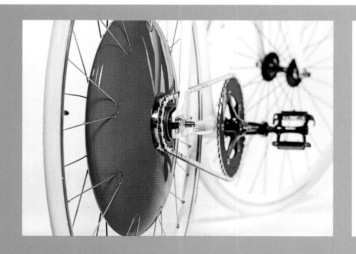

The CERV bike >

The Cannondale CERV concept bike from 2012 adjusted to its rider's needs as it was being pedalled along the road. When riding uphill, the cockpit handlebars adjusted to give a more upright position. As the rider went downhill and speed increased, the bars dropped, to give a more aerodynamic position. Rather than steer through a headset and forks, the bike's direction was dictated via a swing-arm arrangement.

BELOW The Cannondale CERV concept bike, with diagrams alongside. These show the different riding positions that can be made to help tackle different terrain.

85mm

100mm

Climb

Flat

Descend

OPPOSITE LEFT Close-up of the Copenhagen wheel. Inside the enormous red hub, a special brake harnesses energy from the bike's braking and motion. This energy is then transferred to an electric motor, also in the red hub, to augment the rider's pedal power, if it is required.

OPPOSITE RIGHT Loopwheels incorporate suspension within the wheel; a boon to users of bikes with small wheels, which are very rigid. To provide some comfort, small-wheeled bikes may use suspension, although this adds to bike weight and cost.

RIGHT John Villareal's bike of the future is just one designer's concept of what the future of bicycle design may hold. Its futuristic design is defined by a no-chain drive-train and hubless wheels.

Lighting the Way >

In many countries all vehicles on the road have a legal duty to be well-lit at night, a requirement that applies just as much to cyclists as any other road user.

Since the advent of powerful Light Emitting Diodes (LEDs; see pages 38–9) in the 1990s, bike lights have improved massively. Today, simple, inexpensive and compact lights using standard batteries can emit beams as powerful as those from car headlights. On modern roads filled with fast-moving traffic, flashing LED lights help grab the driver's attention and identify the cyclist as a person, rather than just another object on the road.

Brainy Bike Lights >
Brainy Bike Lights comprise a 71-mm (2.8-in) squared module with a motif of a bike on a black background backlit by five LEDs.

A white bike motif is attached to the front of the bike, and a red motif is fixed at the rear. Even when viewed from an acute angle, the motif identifying the vehicle as a bike remains sharp.

It is a simple idea that stands out against the urban clutter of other lights and indicates that 'this is a cyclist'. Tests prove it works; you can even take a reaction-time test to the Brainy Bike Lights on the company's website, www.brainybikelights.com.

Revolights >
Revolights are rings of LEDs that attach to a bike's wheels. They are supplied in sets of four: two white Revolights for the front wheel, and two red for the rear. These lights are powered by lithium ion batteries attached to the wheel hubs. A fork-and-seat-stay-mounted magnet and accelerometer provide speed and orientation data to the light rings so they illuminate only

RIGHT/BELOW Brainy Bike Lights show a cyclist motif at the front and rear of the bike. This announces clearly to other road users that 'this is a bike', so they can identify it immediately, anticipate how it is likely to be ridden on the road and react to it accordingly.

when positioned at the front and rear of a bike. The lights are powerful enough to light the path of the rider while also identifying the vehicle as a cyclist to other road users.

Cycling jacket of the future >

A cycling jacket that incorporates a multi-LED video display over the upper body and sleeves is being developed in America. If successful, it will pave the way for wearable and interactive bicycle lighting, providing a high-tech alternative to the high-visibility yellow tabards some cyclists wear. The technology could see moving images, such as large flashing arrows when turning, being displayed to other road users.

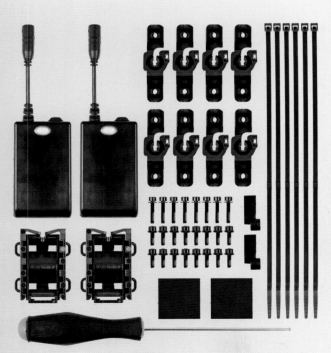

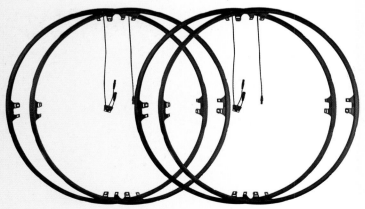

ABOVE/LEFT/BELOW Revolights are another invention that immediately announces that the user is a cyclist. Rings of LEDs on the wheel rims help to identify the bike to other road users, and also illuminate the road ahead for the cyclist.

Stop Thief! >

A good lock is the first step in thief-proofing your bike. Today, there are many highly effective models to choose from with ever more imaginative ways to prevent bike loss.

The bike hoist >

For a TV advert, the German company Conrad made a concept lock that utilized a portable robot in its design. The robot was first attached to a convenient lamppost and then the bike was attached to the robot, which promptly climbed up and out of reach of would-be cycle thieves!

The Skylock >

More than just a concept, the Skylock, available from 2015, heralds the arrival of sophisticated electronics in the area of cycle locks. Powered by solar cells, the Skylock can be locked and unlocked remotely using a mobile phone. If the device is tampered with, a message is sent to the owner's phone alerting them of the fact. In the event of an accident by the bike's user, the Skylock will alert other parties whose details have been pre-programmed in.

BELOW/RIGHT Intelligent bike locks, such as the solar-powered Skylock, will increasingly become a part of bike security. This sophisicated locking system can be operated remotely via the user's phone.

Datatag >

Datatag involves placing hidden identifiers on a bike. They cannot be removed, and they are revealed only when the bike is examined with ultra-violet light. In addition, visible tamper-evident tags are placed on the bike warning off potential thieves. Datatag keep a 24/7 accessible database of all bikes that carry the tags or markings, so if a bike is examined by the authorities its ownership can be established immediately.

Fingerprint locks >

Mobile phone apps with fingerprint recognition technology have been available since 2011 and it will not be long before this technology finds its way – either via a phone app or thumb sensor directly on a keypad – into bike security. It is possible, even likely, that bikes of the future will have some sort of integral fingerprint-activated lock that allows the bike to dock with a network of well-designed bike racks.

LEFT Datatag is a discreet way of tagging any bike with a unique number. The printed number, which is invisible in normal light, is stored in a computer database with the identity and contact details of the bike's owner. If the bike is stolen and then recovered, examination under ultra-violet light reveals the number and, consequently, the identity of the owner.

My Future-Bike >

Ten years ago, I was asked to consider what everyday bikes of the future might look like, and this is what I came up with. Despite its futuristic look, all the elements that make up the bike are already in existence — they have just not been used together to make a single cohesive bike.

Puncture-proof tyres comprise foam-like cells sitting on rims that rotate around stationary 'hoops' on low-friction bearings. On the rim's inner edge are some teeth that engage with the drive system within the frame.

As well as a charging port, the integrated battery is charged via solar cells embedded all over the surface of the bike. The power from this and from braking can be used to supplement pedalling effort on demand via a button on the bars.

Handlebars feature a tiny updatable computer that measures power produced via internal strain gauges and, from this, calculates calories burned. The same display gives satnav information. Also indicated is how much charge is left on the internally stored battery.

A fingerprint-reading sensor, also located on the handlebars, locks and unlocks the wheels via internal pins that are pushed into the rims. This makes it impossible for anyone other than the owner to ride the bike away.

Powerful Light Emitting Diodes (LEDs) are also built into the surface of the bike and run off the same integrated solar-powered battery as the electric drive system.

Future Racers >

Racing is a global sport attracting huge international interest, meaning professional and national teams have to constantly push scientific knowledge and technology if they want to remain competitive. This combined push has revolutionized the way top cyclists train, and caused rapid development in race bikes.

It is inevitable that racing bikes will become even lighter in the future; the fact that they aren't already is due only to cycling's governing body, the UCI, who have imposed a weight limit on any bike used in sanctioned competition. This artificial limit is constantly being challenged and it's believed that the 6.8kg cap will soon be removed altogether, to be replaced instead with a safety standard. Since the cap was introduced in 2000, manufacturers continued to develop and build safe, super-light bikes for serious sports riders who remain unfettered by UCI rules. Pro riders using these machines are forced to add lead weights to their machines to comply with the regulations.

Similarly, disc brakes, a superior braking system to the rim brake developed for the mountain bike world, are outlawed by the UCI but are fast becoming the standard for cyclo-cross and road bikes alike. Despite the disc brake's small weight penalty, its inferior aerodynamics and the fact that it takes more time to change a wheel, it superior performance means it will almost certainly make its way into the professional peloton. The only question is when.

Wireless technology is already used for bike computer and power meters, so shifting gear wirelessly is the logical next step. It is likely more and more performance parameters will be measured and used to refine training and preparation, all built into a new generation of wearable technology.

In the future road-race and time-trial bikes will become ever more aerodynamic. The use of wind tunnels will become more widespread leading to aerodynamically refined riding positions, clothing and helmets. Intelligent fabrics containing micro-pores, which are capable of changing their properties according to the temperature, will revolutionize race clothing. All of these technologies already exist today; they just have not yet found their way into the world of cycling.

Mountain bikes >
Perhaps more than anywhere else, the potential for technological progression in mountain biking is massive. In addition, the off-road

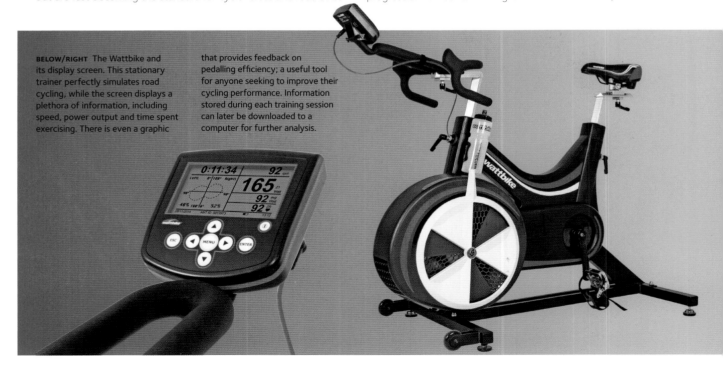

BELOW/RIGHT The Wattbike and its display screen. This stationary trainer perfectly simulates road cycling, while the screen displays a plethora of information, including speed, power output and time spent exercising. There is even a graphic that provides feedback on pedalling efficiency; a useful tool for anyone seeking to improve their cycling performance. Information stored during each training session can later be downloaded to a computer for further analysis.

racing scene is relatively unfettered by regulation, leaving the way clear for the imaginative inventor.

The technology is already available for downhill mountain bikes to have suspension systems that can be adjusted mid-ride – as is possible in motorsport now. Even active-suspension that operates without the need for rider input could find its way into future mountain bikes. Tyres with the capacity to adjust pressure on the fly is a function that would benefit everyone and again is a technology already used in the motor industry.

The majority of future mountain bikes will probably have 27.5in wheels. Mountain bikes with no suspension and fat tyres, which were first developed for riding on snow and sand, are gaining favour as a fun alternative for regular off-road riders. Fat-tyred bikes, as they are called, could even usher in a whole new facet of mountain bike racing, perhaps on sand-based courses in the summer and in winter, on snow.

Single chain-ring gear set-ups, with no gears to be clogged by mud, lend themselves to off-road use and could become a popular trend. At the other end of the scale, improvements in chain technology could see an increase in sprocket number to 12. It is even possible that the current gear system will be married with sophisticated hub gears, giving a near infinite range of available gears to the rider of tomorrow.

Race coverage >

The way we watch cycling is already undergoing massive change. On-bike cameras are taking viewers right into the heart of the event, allowing them to experience something of what it's like to be in a breakaway or in the peloton for the frantic last kilometres of a stage. The same technology is currently being trialled in track racing, adding another facet to the exciting coverage. When technology advances enough to stream these images real-time, allowing them to be incorporated into live coverage (it is currently recorded and downloaded post race) this type of imagery will revolutionize cycling coverage.

In addition, drone-mounted cameras are becoming common in sports coverage, relaying live pictures to TV directors who can switch between drone, rider view, motor bike and helicopter cameras to provide viewers with a perfect picture not only of what a race looks like, but what it feels like.

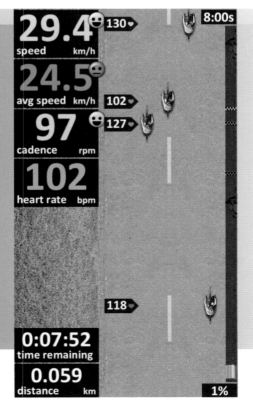

Virtual racing >

Apps such as Race My Ghost already display your current performance on a turbo trainer with your previous best. There's an app for iPhone and Android devices that informs users when they reach Strava segments and provides real-time information on their performance and ranking on each.

It is also possible to register performances made on a Wattbike – a sophisticated brand of home trainer – into a worldwide table, to see where you rank in your gender, age group and preferred distances.

Strava also allows people all over the world to compare their performances in different parameters, such as distance ridden and amount of climbing. If these technologies were linked together in real-time, which surely can't be far away, then the start of a worldwide racing scene, with people competing in gyms and from home, can't be far away either.

LEFT A tablet screen showing the Race My Ghost app. As well as various perfomance measurements, such as speed and heart rate, a graphic shows how far the rider is up or down on previous performances.

Final Word >

'Every time I see an adult on a bicycle I no longer despair for the future of the human race,' the author H G Wells said, early in the 20th century. Maybe he, too, felt that since cycling was something people did as children, those who rode bikes in adulthood still had a child's optimistic outlook and spirit. Now though, in the 21st century, cycling's role in our future is as much practical as it is spiritual.

Just a few years ago, people who rode to work in the UK were seen as smelly, slightly geeky individuals. Now, telling people you ride to the office is actually a small boast. I'm thrilled about how cycling has changed, how it is again being seen as a legitimate form of transport for adults, for normal people, in normal clothes.

I've worked with British Cycling, members of parliament, the All-Party Parliamentary Cycling Group and many more bodies and organizations who can influence our transport choices because I want bikes to become a viable transport choice for people in the UK. I want my kids to be able to ride to school and to the park. I'd like to be able to pedal to the railway station or to the shops. When written down, it doesn't seem a lot to ask. But it can only happen if people are given a safe space in which to ride, if cycling is made an easy choice because human beings are programmed to take the easiest solution. Right now, that is the car. The key is for governments, both local and national, to have a clear and detailed holistic view about what they want their cities to look like in 10 years' time.

Cycling can help our villages, towns and cities to move better. It can improve people's health and help to reduce pollution. These important advantages don't just benefit those who want to ride but everyone in our society.

Bikes today >

Modern bikes are easier to ride than they ever have been at any point in their history. The story is one of continuous improvement while maintaining basic simplicity. There is such a huge range of bikes now; almost anyone can find one to do almost anything on, from riding short commutes to great adventures or breaking new ground in sport. The bike's story has been one of constant innovation and I have no doubt that the story will continue.

Bike design will break new ground, it will improve sporting performance, provide transportation on our urban streets and cost-effective solutions for carrying goods in less developed parts of the world. In the future, designs will enhance durability, reduce the need for maintenance and even help save lives.

This wonderful, versatile, beautiful machine has made its mark on the world, and I have no doubt it will continue to do so for many years to come. I hope we have told the first part of its story well in this book and I look forward to seeing where this two-wheeled marvel travels next.

OPPOSITE Bicycle commuters crossing London Bridge. When city-dwellers trade in their cars for bikes they make more space for everyone on the roads, resulting in faster traffic flow and fewer hold-ups for everyone.

ABOVE Cycling is an excellent way to explore the countryside, and cycle trails like this one are perfect for families wishing to spend leisure time together. Cyclists cover more ground than walkers, meaning they can see more in the same time period.

ABOVE Cycling isn't just practical and efficient. It's not just a way to save the planet and get fit. Cycling is fun, exhilarating, exciting and a great way to refresh and stimulate the senses.

Index >

Picture Credits >

8 Freight 213 below left; Bikefix/Mikołaj Dubisz 213 below right. **50cycles** Focus 250; Kalkhoff 259 left, 259 above right. **2014 Superpedestrian Inc.** Michael D Spencer 270 left. **Adventure Cycling Association** Lon Haldeman 131 above, 131 centre, 131 below. **Alamy** Andrew Lloyd 181 left; ANP 83 above left, 83 above right; blickwinkel 71 below left; Ian Maybury 50 left; inga spence 161 below right; INTERFOTO 32 below; Kathy deWitt 160 right; Konrad Wothe/LOOK Die Bildagentur der Fotografen GmbH 66 left; M&N 61 left; Marjorie Kamys Cotera/Bob Daemmrich Photography 267 left; Mary Evans Picture Library 13 below right; Michael Flippo 68 left; NRT-Sports 162; Old Visuals 66 centre; Petr Bonek 120 below; Photempor 77 right; Photos 12 11; Steve Boyle/NewSport/ZUMAPRESS.com 156 above, 156 below; The Art Archive 13 above right, 25 centre; wanderworldimages 38 above left; Westend61 GmbH 4. **All Star Picture Library** 200. **Art Huneke** 85 below. **Bas de Meijer/HH** 82 above, 82 below left, 82 below right. **BATTAGLIN CICLI SRL** 172. **BBB Cycling** 39, 95. **Bianchi** 228. **Brainy Bike Lights Ltd.** 272 left, 272 right. **Brompton** 115 above left, 115 centre left, 115 above right. **Calfee Design** 43 above left, 43 below left, 43 below right. **Camelbak** 65 above. **Campagnolo S.r.l.** 55 above, 55 below right, 229 below left, 255 above right, 255 centre right, 255 below right. **Cannondale** 67 below, 137 below left, 140, 183, 185 below, 221 right. **Cardboard Technologies** Ilan Besor 43 above. **Chas Roberts/Neil Carison** 139 above. **Classic Cycle Museum** Jeff Groman 98. **Cleland Cycles** Geoff Apps 138, 139 below. **Cor Vos Fotopersburo-Video ENG** HR/Cor Vos © 2013 35 below right. **Corbis** Bettmann 105 centre; Christian Liewig/Liewig Media Sports 125; Colorsport 110; H. Armstrong Roberts/ClassicStock 68 right; Imaginechina 104 above left; New Press Photo/Splash News 53 above; Seattle Post-Intelligencer Collection/Museum of History and Industry 71 above. **Cyclefit** 238 centre above, 238 below. **Datatag** 275. disraeligears.co.uk Michael Sweatman 45 below, 126 left, 126 right. **Dolan Bikes Ltd.** 187 above. **Dynamic Bicycles** 269. **Enigma Bicycle Works** 186 left. **Evans Cycles** 158, 220, 221 above left, 221 below left; BMC 261 above; Scott 231 centre. **Faraday Bikes** 223 below. **Felt Bicycles** 65 below, 101 below, 191 above. **Fred Rompelberg** 84. **Getty Images** A. R. Coster 8; AFP 2, 35 below left, 169, 173 left; Andrew Francis Wallace/Toronto Star 246; Andrey Artykov 97 above right; Branger 26 right; Branger/Roger Viollet 72 below left; Bryn Lennon 6, 244; Carl Court/AFP 105 above; Christian Pondella 157 left; Christophe Ena - IOPP Pool 198; Clive Rose 63 below; Cooksey/Express 99 above right; David Cannon 7 above; David Ramos 128; DEA/G. DAGLI ORTI 14; Duane Howell/The Denver Post 134; Gary M. Prior/Allsport 188, 193 right, 205, 206; Harry How 157 right; Haywood Magee/Picture Post 103 below; Heritage Images 26 left; Hulton Archive 31 right; Imagno 17; J.B. Spector/Museum of Science and Industry, Chicago 28; James Looker/Procycling Magazine 166 below; Jeff Pachoud/AFP 64 right; Joby Sessions/Procycling Magazine 178; Keystone-France 18 right; Keystone-France/Gamma-Keystone 33 above, 33 below, 80, 81 below, 86, 96, 197 below left; Mark Kolbe 58; Mike Ehrmann 61 right; Nicolas Asfouri/AFP 161 below left; Otis Imboden/National Geographic 83 below; Philippe Le Tellier/Paris Match 106; Popperfoto 31 left; Print Collector 10 right; Reinhold Thiele/Henry Guttmann 32 above; Remi Benali/Gamma-Rapho 72 above; Richard Bord 85 above; Robin MacDougall 280; SSPL 25 right, 69 above, 77 left; Stephane De Sakutin/AFP 67 above; STF/AFP 46; Topical Press Agency 72 below right, 160 left; Valery Hache/AFP 194 left; Westend61 97 below right; William Sumits/The LIFE Picture Collection 103 above. **Giant Bicycles/www.giant-bicycles.com** 51 above, 111, 212, 213 above, 226. **Group Lotus plc** 191 below right, 201 centre right, 201 above right, 201 above left, 201 below. **Fuji Bikes/Advanced Sports**

International 161 above. **Hilary Stone** 44 right, 52 left, 52 right, 90, 116 left, 116 right, 122 above left, 122 above right, 122 below left, 122 below right, 123 left, 123 right, 174. **Iban** (CC BY SA 2) 182. **J&R Bicycles** 154. **Janni Turunen** (CC-BY-SA-2) 207 right. **Jelly Products Ltd.** 270 right. **Joe Breeze** 135 below. **John Villarreal** 271 below. **J P Weigle** 115 below left, 119 centre left, 119 below left, 119 below right. **Kona Bicycle Company** 136, 231 above. **LOOK – JP Ehrmann** 175 left, 184 right. **Madison** Continental 27 above right, 27 centre right; Garmin 180 above right, 180 below right, 209 below left, 256, 257 left; Genesis 215 below left, 215 below right; GoPro 257 right; Shimano 22, 54, 56 below, 145 below right, 175 centre, 175 right, 254, 255 above left, 255 centre left, 255 centre, 255 below left. **Magura** 145 above left, 145 above right. **Mary Evans Picture Library** Illustrated London News Ltd. 18 left. **Mavic** 121 below right, 166 above, 252; Christophe Margot 19 below; Jeremie Reuiller 147 above left, 147 above right, 167 above, 179 above left, 179 above right; Loris von Siebenthal 235 above; Nicolas Jutzi 229 above left; Pierre Thomas 19 above, 146. **Moebiusuibeom-en** (CC-by-SA 3.0) 23 below. **Morewood Bikes** 143 above left. **Mosquito Bikes** Dario Pergoretti 118. **Moulton** 78, 108, 109 above. **National Clarion Cycling Club** 30 above, 30 below, 31 centre. **Offside** Archivi Farabola 93 above; FARABOLAFOTO 181 right; Gerry Cranham 63 above; IPP 241 above; L'Equipe 27 below left, 35 above right, 40 left, 40 right, 48, 51 below, 55 below left, 64 below left, 76 left, 76 right, 88 left, 88 right, 89, 93 centre, 127 above, 127 below, 132, 133 left, 168, 170, 173 right, 187 below, 190, 193 below left, 194 right, 195, 196, 197 above right, 197 centre right above, 197 centre right below, 197 below right, 202, 203 left, 207 left, 216, 217 left, 229 below right, 261 below; Pressesports 49, 93 below, 133 right, 203 right, 204; Rob Lampard 217 right. **Paligap** Ritchey 137 right; Saris Cycling Group 209 above left, 209 above centre, 253 left, 253 right.

Peter Lawrence Lucas 20, 38 below right. **PhotoSport International** Ron Good 164 right. **Priority Designs** 264, 271 above. **Qhubeka** 263 left. **Raleigh UK Ltd.** 41, 101 above, 113 above right, 113 above left, 164 left, 165; Dahon 114. **Red Bull Content Pool** Garth Milan 218, 231 below. **Retül** Allen Krughoff 239. **Reuters** Luc Novovitch LN/CMC 167 below. **Revolights** 273 above, 273 below left, 273 below right. **Rex Features** Colorsport 191 below left; John Pierce Owner PhotoSport Int 192. **Riese & Müller GmbH** 109 below left, 109 below right. **Rohloff AG** 57 below left, 57 below right. **Scott Sports SA** 241 below. **Shutterstock** africa924 263 right; ChameleonsEye 81 above; Dutourdumonde Photography 105 below; Fifian Iromi 74; FomaA 104 right; Gritsana P 267 right; Hans Engbers 79 right; Iakov Filimonov 97 left; Monkey Business Images 281 left; Pichi 44 left; Susan Leggett 281 right; T.W. van Urk 79 left; Tupungato 104 below left, 266 left; urosr 71 below right. **Skylock** 274 above, 274 below. **Specialized** 135 above left, 143 centre left, 143 below, 147 below, 149 above left, 150, 223 above, 234 above. **Speedplay, Inc.** 176, 179 below, 180 left, 184 left, 186 right, 193 above left. **SRAM** 57 above, 140 above left, 143 above centre, 143 above right, 144, 145 below left, 232, 234 below. **SRM GmbH** 208. **Stages Cycling** 209 below right. **Sturmey-Archer Europa N.V.** 56 above. **The Vanilla Workshop** (CC BY-ND 2.0) 119 above right. **timquijano** (CC BY 2.0) 121 above right. **Trek Bicycle Corporation** 142 below right, 185 above, 238 centre below, 268; Chris Bacarella 238 above; Jamie Forrest 229 above right. **Trevor Jarvis** 91 centre. **Turbojams** (CC-BY-SA 2.0) via Flikr 100. **University of Montana**, U.M. Collection [73.0031], F.M Ingalls, Archives & Special Collections, Mansfield Library 130. **Vanilla Bikes** Bob Huff Photo 119 above left. **Velocomp LLC** 209 above right. **velosolo.co.uk** 38 above right, 73. **Via Wikipedia** 15, 50. **Wattbike Ltd.** 278. **Wende Cragg** 135 above right. **www.klausmaiphotographie.de** 240.

Acknowledgements >

Octopus Publishing Group would like to thank **Chris Boardman** and **Chris Sidwells** for their hard work and expertise in producing this book.

Thanks to **Peter Dawson** of Grade Design for the layout and **Tif Hunter** for the special photography. Thanks also to **Jamie Mitchell** and **James Ryan** of Boardman Bikes for their assistance with the images in the later chapters. We'd also like to thank **John Gill** of the National Cycle Collection, Mid Wales, for his help in facilitating the photography at The National Cycle Collection museum. The museum is run by volunteers and trustees and is open 10.00am–4.00pm on Tuesdays, Wednesdays and Fridays. For more information, visit http://www.cyclemuseum.org.uk.